LONDON MATHEMATICAL SOCIETY LECTURE NOT

Managing Editor: Professor Endre Süli, Mathematical Institute, University of C
Woodstock Road, Oxford OX2 6GG, United Kingdom

The titles below are available from booksellers, or from Cambridge University Press at
www.cambridge.org/mathematics

London Mathematical Society Lecture Note Series: 460

Wigner-Type Theorems
for Hilbert Grassmannians

MARK PANKOV

University of Warmia and Mazury in Olsztyn, Poland

CAMBRIDGE
UNIVERSITY PRESS

CAMBRIDGE
UNIVERSITY PRESS

University Printing House, Cambridge CB2 8BS, United Kingdom

One Liberty Plaza, 20th Floor, New York, NY 10006, USA

477 Williamstown Road, Port Melbourne, VIC 3207, Australia

314–321, 3rd Floor, Plot 3, Splendor Forum, Jasola District Centre,
New Delhi – 110025, India

79 Anson Road, #06–04/06, Singapore 079906

Cambridge University Press is part of the University of Cambridge.

It furthers the University's mission by disseminating knowledge in the pursuit of
education, learning, and research at the highest international levels of excellence.

www.cambridge.org
Information on this title: www.cambridge.org/9781108790918
DOI: 10.1017/9781108800327

© Cambridge University Press 2020

This publication is in copyright. Subject to statutory exception
and to the provisions of relevant collective licensing agreements,
no reproduction of any part may take place without the written
permission of Cambridge University Press.

First published 2020

Printed and bound in Great Britain by Clays Ltd, Elcograf S.p.A.

A catalogue record for this publication is available from the British Library.

ISBN 978-1-108-79091-8 Paperback

Cambridge University Press has no responsibility for the persistence or accuracy of
URLs for external or third-party internet websites referred to in this publication
and does not guarantee that any content on such websites is, or will remain,
accurate or appropriate.

Contents

Preface

Wigner's theorem [67] provides a geometric characterization of unitary and anti-unitary operators as transformations of the set of rays of a complex Hilbert space or, equivalently, rank one projections. This statement plays an important role in the mathematical foundations of quantum mechanics [11, 50, 63], since rays (rank one projections) can be identified with pure states of quantum mechanical systems. We present various types of extensions of Wigner's theorem onto Hilbert Grassmannians and their applications. Most of these results were obtained after 2000, but for completeness of the exposition we include some classic theorems closely connected to the main topic (for example, Kakutani and Mackey's result on the lattice of closed subspaces of a complex Banach space [31] and Kadison's theorem on transformations preserving the convex structure of the set of states of quantum mechanical systems [30]). We use geometric methods related to the Fundamental Theorem of Projective Geometry and results in the spirit of Chow's theorem [13].

Formally, the material is based on the first part of Varadarajan's book [63, Chapters I – IV], but we do not assume that readers are familiar with this book. So, we describe briefly all relevant facts from the mathematical foundations of quantum mechanics and refer to Birkhoff and von Neumann [6] for motivations, physical background and the detailed description. Since the requirement of readers is only knowledge of the basics of linear algebra and operator theory, the book is accessible for graduate students.

I am very grateful to Antonio Pasini for useful discussions on some parts of this book. Finally, I would like to express my sincere gratitude to the anonymous reviewer whose remarks, suggestions and corrections helped me to finish the book.

Mark Pankov

Introduction

It was first observed by Birkhoff and von Neumann [6] that the logical structure of quantum mechanics is related to the orthomodular lattice formed by closed subspaces of a complex Hilbert space. On each orthomodular lattice is defined an important class of functions called *states*; all states form a convex set whose extreme points are known as *pure states* [63, Section III.3]. Gleason's theorem [26] describes the set of states for the orthomodular lattices associated to separable complex Hilbert spaces. It says that all states can be identified with bounded self-adjoint positive operators of trace one; in particular, pure states correspond to rank one projections, i.e. rays of the Hilbert space.

The classic Wigner's theorem [67] (see also [11, 50, 63]) characterizes unitary and anti-unitary operators as symmetries of quantum mechanical systems, i.e. every bijective transformation of the set of pure states preserving the transition probability is induced by a unitary or anti-unitary operator. We refer to Chevalier [12] for a history and a brief description of the physical background (for example, it is shown how to derive the Schrödinger equation for a conservative physical system from Wigner's theorem).

In this book, readers will meet two versions of Wigner's theorem. The non-bijective version says that an arbitrary transformation of the Grassmannian formed by rays of a complex Hilbert space which preserves the angles between pairs of rays (the square of the cosine of such an angle is equal to the transition probability between the corresponding pure states) is induced by a linear or conjugate-linear isometry. On the other hand, it was observed by Uhlhorn [62] that to get a unitary or anti-unitary operator it is sufficient to require that a transformation of the Grassmannian of rays is a bijection preserving the orthogonality relation in both directions and the dimension of the Hilbert space is not less than three. Note that the non-bijective analogue of the latter statement does not hold for infinite-dimensional Hilbert spaces. Uhlhorn's theorem is a simple consequence of the Fundamental Theorem of Projective Geometry (for this reason, the dimension of the Hilbert space is assumed to be not less than three); but it reveals the following important relation between the logical structure and the probabilistic structure of quantum mechanical systems: if the logical structure is preserved, then probabilistic structure also is preserved.

The description of bijective transformations preserving the convex structure of the set of all quantum states (the set of all bounded self-adjoint positive operators of trace one) [30] is a classic application of Uhlhorn's version of Wigner's theorem. Since pure states are characterized as extreme points of the convex set of all states, every such transformation induces a bijective transformation of the set of pure states. The latter transformation preserves the orthogonality relation in both directions (this fact is non-trivial) and we come to a unitary or anti-unitary operator.

We present Wigner type theorems for Hilbert Grassmannians. It must be pointed out that we distinguish the Grassmannians whose elements are finite-dimensional subspaces (the dual objects are the Grassmannians consisting of closed subspaces of finite codimensions) and the Grassmannians formed by closed subspaces whose dimension and codimension both are infinite. Results of such a kind were first obtained in [36] and [27, 59], where the non-bijective and Uhlhorn's versions of Wigner's theorem were extended on other Grassmannians. Molnár's theorem [36] states that transformations of Grassmannians (formed by finite-dimensional subspaces) preserving the principal angles between any pair of subspaces are induced by linear and conjugate-linear isometries (except one finite-dimensional case). We generalize this result and show that it is sufficient to assume that only some types of the principal angles are preserved. Another generalization of Molnár's theorem was proved by Gehér [25].

Györy [27] and Šemrl [59] (independently) described bijective transformations of Hilbert Grassmannians preserving the orthogonality relation in both directions. Note that a non-bijective version of this result holds only for finite-dimensional Hilbert spaces. One of the applications of the Györy–Šemrl theorem is the determination of isometries of Hilbert Grassmannians with respect to the gap metric [24].

In the case when the Grassmannian consists of closed subspaces whose dimension and codimension both are infinite, a bijective transformation preserving the orthogonality relation (in both directions) is also inclusions preserving and we show that it can be extended to an automorphism of the lattice of closed subspaces. It is well known that all automorphisms of the lattice of closed subspaces of an infinite-dimensional complex normed space are induced by linear and conjugate-linear homeomorphisms of the normed space to itself (for the finite-dimensional case this fails). This fact was established by

Kakutani and Mackey [31] as a step in the proof of the following remarkable result: every orthomodular lattice formed by all closed subspaces of an infinite-dimensional complex Banach space is the orthomodular lattice associated to a complex Hilbert space.

We also investigate compatibility preserving transformations. The compatibility relation is one of the basic concepts of quantum logic. The orthomodular lattice formed by closed subspaces of a complex Hilbert space is considered as the standard quantum logic. Elements of this lattice are identified with projections, i.e. self-adjoint idempotents in the Banach algebra of bounded operators. Two closed subspaces are compatible if and only if the corresponding projections commute. Two distinct rays are compatible only in the case when they are orthogonal. For this reason, we regard statements which describe compatible preserving transformations as Wigner type theorems.

We will use geometric methods based on properties of Grassmann graphs in the spirit of [15, 45]. So, the Fundamental Theorem of Projective Geometry, Chow's theorem [13], apartments and their orthogonal analogues will be useful tools for our investigations. We include a chapter on geometric transformations of Grassmannians associated to vector spaces of arbitrary (not necessarily finite) dimension. A large portion of the results of this chapter is new and cannot be found in [15, 44, 45].

At the end, we give a few words on applications. It was noted above that Uhlhorn's version of Wigner's theorem was exploited in determining bijective transformations preserving the convex structure of the set of all quantum states. In a similar way, we will use analogues of Wigner's theorem for Hilbert Grassmannians to study linear transformations of the real vector space of self-adjoint finite-rank operators which send projections of fixed rank to projections of the same rank [1, 57, 58] or to projections of other fixed rank [49].

1

Two Lattices

We describe briefly some basic properties of the lattice formed by all sub-spaces of a vector space and the orthomodular lattice consisting of all closed subspaces of a complex Hilbert space. The first lattice is investigated in classic projective geometry [3]. The second is related to the logical structure of quantum mechanical systems (we refer to [19, 63] for the details and strongly recommend the short problem book [14] as a quick introduction to the topic).

1.1 Lattices

Let X be a non-empty set with a certain relation denoted by $\leq$. The pair $(X, \leq)$ is called a *partially ordered set* if for all $x, y, z \in X$ the following three conditions hold:

- $x \leq x$;
- if $x \leq y$ and $y \leq x$, then $x = y$;
- if $x \leq y$ and $y \leq z$, then $x \leq z$.

A partially ordered set $(X, \leq)$ is said to be a *lattice* if it satisfies the following additional conditions:

- for any two elements $x, y \in X$ there is the *least upper bound* $x \vee y$, i.e. an element $z \in X$ such that $x \leq z$, $y \leq z$ and we have $z \leq z'$ for all $z' \in X$ satisfying $x \leq z'$ and $y \leq z'$;
- for any two elements $x, y \in X$ there is the *greatest lower bound* $x \wedge y$, i.e. an element $t \in X$ such that $t \leq x$, $t \leq y$ and we have $t' \leq t$ for all $t' \in X$ satisfying $t' \leq x$ and $t' \leq y$.

A lattice is called *bounded* if it contains the *least* element 0 and the *greatest* element 1 such that $0 \leq x \leq 1$ for every element x. A lattice $(X, \leq)$ is *complete* if for every subset $Y \subset X$ there is the least upper bound $\bigvee_{y \in Y} y$ and the greatest lower bound $\bigwedge_{y \in Y} y$.

An *isomorphism* between partially ordered sets $(X, \leq)$ and $(X', \leq)$ is a bijection $f : X \to X'$ preserving the order $\leq$ in both directions, i.e. for $x, y \in X$ we have

$$x \leq y \iff f(x) \leq f(y).$$

If these partially ordered sets are lattices and $f : X \to X'$ is an isomorphism between them, then

$$f(x \vee y) = f(x) \vee f(y) \text{ and } f(x \wedge y) = f(x) \wedge f(x)$$

for all $x, y \in X$; moreover, if our lattices are complete, then

$$f\left(\bigvee_{y \in Y} y\right) = \bigvee_{y \in Y} f(y) \text{ and } f\left(\bigwedge_{y \in Y} y\right) = \bigwedge_{y \in Y} f(y)$$

for any subset $Y \subset X$. Isomorphisms of bounded lattices transfer the least and greatest elements to the least and greatest elements, respectively.

A bijection $g : X \to X'$ is said to be an *anti-isomorphism* of $(X, \leq)$ to $(X', \leq)$ if it is order reversing in both directions, i.e.

$$x \leq y \iff g(y) \leq g(x)$$

for all $x, y \in X$. If our partially ordered sets are lattices and $g : X \to X'$ is an anti-isomorphism between them, then

$$g(x \vee y) = g(x) \wedge g(y) \text{ and } g(x \wedge y) = g(x) \vee g(y)$$

for all $x, y \in X$; also, we have

$$g\left(\bigvee_{y \in Y} Y\right) = \bigwedge_{y \in Y} g(y) \text{ and } g\left(\bigwedge_{y \in Y} Y\right) = \bigvee_{y \in Y} g(y)$$

for any subset $Y \subset X$ if our lattices are complete. Anti-isomorphisms of bounded lattices transpose the least and greatest elements.

Example 1.1 For every non-empty set X we denote by $\mathcal{L}(X)$ the set of all subsets of X. The partially ordered set $(\mathcal{L}(X), \subset)$ is a bounded lattice. If A and B are subsets of X, then their least upper bound is $A \cup B$ and their greatest lower bound is $A \cap B$. The least element of the lattice $\mathcal{L}(X)$ is the empty set and the greatest element is X. This lattice is complete. The lattices $\mathcal{L}(X)$ and $\mathcal{L}(Y)$ are isomorphic if and only if the sets X and Y are of the same cardinality. In this case, every isomorphism of these lattices is induced by a bijection between X and Y.

A bounded lattice is said to be *complemented* if for every element x there is a *complement* x', i.e. an element x' satisfying

$$x \wedge x' = 0 \text{ and } x \vee x' = 1.$$

A *Boolean algebra* is a complemented lattice with the following distributive rules:

$$x \vee (y \wedge z) = (x \vee y) \wedge (x \vee z),$$

$$x \wedge (y \vee z) = (x \wedge y) \vee (x \wedge z).$$

Using these rules, we can show that for every element of a Boolean algebra there is the unique complement (see, for example, [63, p. 8]).

Example 1.2   The lattice $\mathcal{L}(X)$ from Example 1.1 is a Boolean algebra.

Let $(X, \leq)$ be a bounded lattice. An *orthocomplementation* is a transformation $x \to x^\perp$ such that for all $x, y \in X$ the following conditions hold:

(1) $x \vee x^\perp = 1$ and $x \wedge x^\perp = 0$,
(2) $x^{\perp\perp} = x$,
(3) if $x \leq y$, then $y^\perp \leq x^\perp$.

The conditions (2) and (3) imply that the orthocomplementation is an anti-automorphism of $(X, \leq)$. Hence $0^\perp = 1$ and $1^\perp = 0$. For elements $x, y \in X$ we write $x \perp y$ and say that these elements are *orthogonal* if $x \leq y^\perp$ (this relation is symmetric, since $x \leq y^\perp$ implies that $y \leq x^\perp$).

A bounded lattice with an orthocomplementation is called *orthomodular* if for any two elements x, y satisfying $x \leq y$ we have

$$x \vee (x^\perp \wedge y) = y.$$

In such a lattice, De Morgan's laws

$$(x \vee y)^\perp = x^\perp \wedge y^\perp \text{ and } (x \wedge y)^\perp = x^\perp \vee y^\perp$$

hold true [63, Lemma 3.1]. Two elements x, y of an orthomodular lattice are said to be *compatible* if

$$x' = x \wedge (x \wedge y)^\perp \text{ and } y' = y \wedge (x \wedge y)^\perp \tag{1.1}$$

are orthogonal. For example, x and y are compatible if $x \leq y$ or $x \perp y$.

Example 1.3   Every Boolean algebra is an orthomodular lattice whose orthocomplementation is the complementation. Let x, y be elements of a Boolean algebra and let x', y' be as in (1.1). Using the distributive rules and De Morgan's laws, we establish that $x' = x \wedge y^\perp$ and $y' = y \wedge x^\perp$, which implies that

$x' \wedge y' = 0$, i.e. x' and y' are orthogonal. Therefore, any two elements in a Boolean algebra are compatible.

Remark 1.4 By [63, Lemma 3.7], two elements in an orthomodular lattice $(X, \leq)$ are compatible if and only if there is a subset $X' \subset X$ containing these elements and such that $(X', \leq)$ is a Boolean algebra.

Let $(X, \leq)$ be an orthomodular lattice such that for every countable subset there is a least upper bound. A function $p : X \rightarrow [0, 1]$ is called a *state* if it satisfies the following conditions:

- $p(0) = 0$ and $p(1) = 1$,
- for every countable subset $\{x_i\}_{i \in I}$ formed by mutually orthogonal elements we have

$$p\left(\bigvee_{i \in I} x_i\right) = \sum_{i \in I} p(x_i).$$

If I is a countable set, $\{p_i\}_{i \in I}$ are states and $\{t_i\}_{i \in I}$ are non-negative real numbers such that $\sum_{i \in I} t_i = 1$, then the function $p : X \rightarrow [0, 1]$ defined as

$$p(x) = \sum_{i \in I} t_i p_i(x) \text{ for all } x \subset X$$

is a state, i.e. the set of all states is convex. Extreme points of this convex set are said to be *pure states*. In other words, a state p is pure if for any states p_1, p_2 and any $t \in (0, 1)$ the equality

$$p = tp_1 + (1 - t)p_2$$

implies that $p = p_1 = p_2$.

Example 1.5 Consider the Boolean algebra $\mathcal{L}(X)$ formed by all subsets of a set X. For $x \in X$ we define $p_x(y) = \delta_x^y$ for every $y \in X$ (δ_x^y is the Kronecker symbol) and extend p_x on $\mathcal{L}(X)$ as follows:

$$p_x(A) = \begin{cases} 1 & \text{if } x \in A, \\ 0 & \text{if } x \notin A; \end{cases}$$

it is clear that p_x is a state. For every state $p : \mathcal{L}(X) \rightarrow [0, 1]$ the set

$$A_p = \{x \in X : p(x) > 0\}$$

is countable (otherwise, there is a natural number $n > 1$ such that the set of all $x \in X$ satisfying $p(x) > 1/n$ is uncountable, which is impossible). If X is countable, then p is completely determined by the values on elements of X, i.e.

$$p = \sum_{x \in A_p} t_x p_x,$$

where each t_x is greater than 0 and $\sum_{x\in A_p} t_x = 1$ (since $p(A_p) = 1$). In this case, p is a pure state if and only if $p = p_x$ for a certain $x \in X$. In the general case, the same holds if and only if X is a set of non-measurable cardinality [17, Chapter 6, Theorem 1.4].

1.2 The Lattice of Subspaces of a Vector Space

Let V be a left vector space over a division ring R, i.e. V is an additive abelian group (whose identity element is denoted by 0) and there is a left action of the division ring R on V satisfying the following conditions:

(1) $1x = x$ for all $x \in V$,
(2) $a(x + y) = ax + ay$ for all $a \in R$ and $x, y \in V$,
(3) $(a + b)x = ax + bx$ for all $a, b \in R$ and $x \in V$,
(4) $a(bx) = (ab)x$ for all $a, b \in R$ and $x \in V$

(using (2) and (3) we show that $a0 = 0$ for every $a \in R$ and $0 \in V$ and $0x = 0$ for every $x \in V$ and $0 \in R$). This action can be considered as a right action of the *opposite division ring* R^*. The division rings R and R^* have the same set of elements and the same additive operation. The multiplicative operation $a*b$ on R^* is defined as $b \cdot a$, where $\cdot$ is the multiplicative operation on R (note that R coincides with R^* in the commutative case). For the corresponding right action of R^* on V the condition (4) is rewritten as

$$(xb)a = x(b * a).$$

Every left or right vector space over R is a right or, respectively, left vector space over R^*.

Denote by $\mathcal{L}(V)$ the set of all subspaces of V. The partially ordered set $(\mathcal{L}(V), \subset)$ is a bounded lattice. For any two subspaces X and Y the least upper bound is $X + Y$ and the greatest lower bound is $X \cap Y$. The least element is 0 and the greatest element is V. This lattice is complete.

If $\dim V = 1$, then the lattice consists of the least element and the greatest element only. In the case when $\dim V = 2$, every element of $\mathcal{L}(V)$ distinct from 0 and V is a 1-dimensional subspace and for any proper subspaces $X, Y \subset V$ the inclusion $X \subset Y$ implies that the subspaces are coincident. For this reason, we will always suppose that $\dim V \geq 3$.

Remark 1.6 A complemented lattice is called *modular* if for any element x and elements y, z satisfying $y \leq z$ we have

$$(x \vee y) \wedge z = (x \wedge z) \vee y.$$

The *rank* of a lattice is the maximal number of non-zero elements in linearly ordered subsets. It is well known that a modular lattice of rank ≥ 4 is the lattice formed by all subspaces of a left vector space over a division ring if for every element there is more than one complement (it must be pointed out that the rank is not assumed to be finite, see [3, Chapter VII]), and we need the additional desarguesian axiom to state the same for the case of rank three.

If B is a basis of the vector space V, then the set $\mathcal{A}$ consisting of all subspaces spanned by subsets of B is said to be the *apartment* of $\mathcal{L}(V)$ associated to the basis B. The partially ordered set $(\mathcal{A}, \subset)$ is a complete Boolean algebra isomorphic to the Boolean algebra formed by all subsets of a set whose cardinality is the dimension of V. Two bases define the same apartment if and only if the vectors from one basis are scalar multiples of the vectors from the other.

Proposition 1.7 *For any two elements of $\mathcal{L}(V)$ there is an apartment containing them.*

Proof For any two subspaces X, Y we take a basis of $X \cap Y$ and extend it to bases of X and Y. The union of these bases is an independent subset and we extend it to a basis of V. The associated apartment contains both X and Y. □

Remark 1.8 If V is finite-dimensional, then the lattice $\mathcal{L}(V)$ together with the family of all apartments is a structure closely connected to the Tits building of the general linear group $GL(V)$ (see [61] for the details).

The *Grassmannians* of the vector space V can be defined as the orbits of the action of the general linear group $GL(V)$ on the lattice $\mathcal{L}(V)$. If V is finite-dimensional, then $\mathcal{G}_k(V)$ is the Grassmannian formed by all k-dimensional subspaces of V, where $1 \leq k \leq \dim V - 1$. Suppose that $\dim V = \alpha$ is an infinite cardinality. For every cardinality $\beta \leq \alpha$ we denote by $\mathcal{G}_\beta(V)$ the Grassmannian consisting of all subspaces $X \subset V$ such that

$$\dim X = \beta \quad \text{and} \quad \operatorname{codim} X = \alpha,$$

and we write $\mathcal{G}^\beta(V)$ for the Grassmannian formed by all subspaces $Y \subset V$ satisfying

$$\dim Y = \alpha \quad \text{and} \quad \operatorname{codim} Y = \beta.$$

Then $\mathcal{G}_\alpha(V) = \mathcal{G}^\alpha(V)$ consists of all subspaces whose dimension and codimension both are α. If β is an infinite cardinality and $\mathcal{G}$ is $\mathcal{G}_\beta(V)$ or $\mathcal{G}^\beta(V)$, then for every $X \in \mathcal{G}$ there are infinitely many elements of $\mathcal{G}$ incident to X; we note that the partially ordered set $(\mathcal{G}, \subset)$ is not a lattice. The intersections of a Grassmannian with apartments of $\mathcal{L}(V)$ will be called *apartments* of this Grassmannian.

The *dual vector space* V^* (formed by all linear functionals on V) is a right vector space over R. We will consider V^* as a left vector space over the opposite division ring R^*.

Let $\{e_i\}_{i \in I}$ be a basis of V. Consider the vectors $\{e_i^*\}_{i \in I}$ in V^* satisfying $e_i^*(e_j) = \delta_j^i$ for any pair $i, j \in I$, where δ_j^i is the Kronecker symbol. These vectors form a linearly independent subset of V^*. If V is finite-dimensional, then this is a basis of V^* and we have $\dim V = \dim V^*$. In the case when V is infinite-dimensional, there are elements of V^* which are non-zero on infinitely many e_i and we get $\dim V < \dim V^*$.

Theorem 1.9 *If* $\dim V = \alpha$ *is infinite, then* $\dim V^* = \beta^\alpha$, *where* β *is the cardinality of the associated division ring*[1].

Proof   See [3, Section II.3].   □

For every subset $X \subset V$ we define the *annihilator* X^0 as the set of all $x^* \in V^*$ satisfying $x^*(x) = 0$ for all $x \in X$. It is clear that X^0 is a subspace in V^*. For every subset $Y \subset V^*$ the (left) *annihilator* ^0Y is the subspace of V formed by all vectors $y \in V$ such that $y^*(y) = 0$ for all $y^* \in Y$.

Remark 1.10   If V is finite-dimensional, then the second dual space V^{**} can be naturally identified with V. Every vector $x \in V$ defines the linear functional $x^* \to x^*(x)$ on V^* and this correspondence is a linear isomorphism of V to V^{**}. Then $^0X = X^0$ for every subspace $X \subset V^*$.

For every subspace $X \subset V$ we have $^0(X^0) = X$, and

$$(X + Y)^0 = X^0 \cap Y^0, \quad (X \cap Y)^0 = X^0 + Y^0$$

for all subspaces $X, Y \subset V$. Similarly, $(^0X')^0 = X'$ for every subspace $X' \subset V^*$ and

$$^0(X' + Y') = {}^0X' \cap {}^0Y', \quad {}^0(X' \cap Y') = {}^0X' + {}^0Y'$$

for all subspaces $X', Y' \subset V^*$.

Denote by $\mathcal{L}_{\mathrm{fin}}(V)$ and $\mathcal{L}^{\mathrm{fin}}(V)$ the sets of all subspaces of finite dimension and finite codimension, respectively. If V is finite-dimensional, then $\mathcal{L}_{\mathrm{fin}}(V)$ and $\mathcal{L}^{\mathrm{fin}}(V)$ both are coincident with $\mathcal{L}(V)$. In the case when V is infinite-dimensional, the partially ordered sets $(\mathcal{L}_{\mathrm{fin}}(V), \subset)$ and $(\mathcal{L}^{\mathrm{fin}}(V), \subset)$ are unbounded lattices. The following facts are well known:

- if $X \in \mathcal{L}^{\mathrm{fin}}(V)$, then $X^0 \in \mathcal{L}_{\mathrm{fin}}(V^*)$ and the dimension of X^0 is equal to the codimension of X;

---

[1] Recall that β^α is the cardinality of the set formed by all maps from a set of cardinality α to a set of cardinality β.

- if $Y \in \mathcal{L}_{\text{fin}}(V^*)$, then $^0Y \in \mathcal{L}^{\text{fin}}(V)$ and the codimension of ^0Y is equal to the dimension of Y.

The annihilator map of $\mathcal{L}^{\text{fin}}(V)$ to $\mathcal{L}_{\text{fin}}(V^*)$ is an anti-isomorphism of these lattices and the inverse anti-isomorphism also is the annihilator map. This anti-isomorphism cannot be extended to an anti-isomorphism between the lattices $\mathcal{L}(V)$ and $\mathcal{L}(V^*)$ if V is infinite-dimensional. If $\dim V = n$ is finite, then the annihilator map is an anti-isomorphism of $\mathcal{L}(V)$ to $\mathcal{L}(V^*)$ which sends every $\mathcal{G}_k(V)$ to $\mathcal{G}_{n-k}(V^*)$ and transfers the apartment defined by a basis $e_1, \dots, e_n$ to the apartment defined by the dual basis $e_1^*, \dots, e_n^*$.

Remark 1.11 Suppose that V is infinite-dimensional and $B = \{e_i\}_{i \in I}$ is a basis of V. The vectors $e_i^* \in V^*$, $i \in I$ satisfying $e_i^*(e_j) = \delta_j^i$ form a linearly independent subset of V^* and we take any basis B' of V^* containing them. The annihilator map transfers the apartment of $\mathcal{G}^k(V)$ associated to B to a proper subset of the apartment of $\mathcal{G}_k(V^*)$ defined by B'. So, the annihilator map sends every apartment of $\mathcal{G}^k(V)$ to a subset contained in infinitely many apartments of $\mathcal{G}_k(V^*)$.

1.3 The Orthomodular Lattice of Closed Subspaces of a Hilbert Space

Recall that a *complex inner product space* is a complex vector space H together with an inner product $H \times H \to \mathbb{C}$; we denote by $\langle x, y \rangle$ the inner product of vectors $x, y \in H$ and require that the following conditions are satisfied:

(1) $\langle y, x \rangle = \overline{\langle x, y \rangle}$ for any $x, y \in H$,
(2) the inner product is linear in the first argument, i.e. for any vectors $x, y, z \in H$ and scalars $a, b \in \mathbb{C}$ we have $\langle ax + by, z \rangle = a\langle x, z \rangle + b\langle y, z \rangle$,
(3) for each vector $x \in H$ the product $\langle x, x \rangle$ is a non-negative real number which is equal to 0 if and only if $x = 0$.

Then (1) and (2) imply that $\langle z, ax + by \rangle = \overline{a}\langle z, x \rangle + \overline{b}\langle z, y \rangle$ for all $x, y, z \in H$ and $a, b \in \mathbb{C}$. For every vector $x \in H$ we define the *norm*

$$\|x\| = \sqrt{\langle x, x \rangle}.$$

The following assertions are fulfilled:

- $\|ax\| = |a| \cdot \|x\|$ for all $x \in H$ and $a \in \mathbb{C}$,
- $\|x + y\| \leq \|x\| + \|y\|$ for all $x, y \in H$,
- $|\langle x, y \rangle| \leq \|x\| \cdot \|y\|$ for all $x, y \in H$.

In particular, an inner product space can be considered as a metric space, where the distance between any two vectors x and y is equal to $\|x - y\|$. An inner product space is called a *Hilbert space* if this metric space is complete, i.e. every Cauchy sequence converges.

Let H be a complex Hilbert space. Denote by $\mathcal{L}(H)$ the set of all closed subspaces of H. The partially ordered set $(\mathcal{L}(H), \subset)$ is a bounded lattice whose least element is 0 and whose greatest element is H. For any two closed subspaces $X, Y \subset H$ the greatest lower bound is the intersection $X \cap Y$ and the least upper bound is $X + Y$, i.e. the minimal closed subspace containing the sum $X + Y$. This lattice is complete. If H is finite-dimensional, then every subspace of H is closed and $\mathcal{L}(H)$ coincides with the lattice considered in the previous section.

Two vectors are called *orthogonal* if their inner product is zero. A vector of norm one is said to be *unit*. Recall that an *orthonormal basis* of the Hilbert space H is a maximal subset formed by mutually orthogonal unit vectors. Any two orthonormal bases are of the same cardinality which is called the *dimension* of the Hilbert space H. A Hilbert space is *separable* if its dimension is countable. If $\{e_i\}_{i \in I}$ is an orthonormal basis of H, then for every vector $x \in H$ there is a countable subset $J(x) \subset I$ such that the inner product $\langle x, e_i \rangle$ is nonzero only for $i \in J(x)$, the series

$$\sum_{i \in J(x)} \langle x, e_i \rangle e_i$$

converges to x (in the norm) and

$$\|x\|^2 = \sum_{i \in J(x)} |\langle x, e_i \rangle|^2.$$

Similarly, the *dimension* of a closed subspace $X \subset H$ is the cardinality of orthonormal bases of X, i.e. maximal subsets of X consisting of mutually orthogonal unit vectors.

Example 1.12   The sum of closed subspaces is not necessarily closed. The following example is taken from [55]. Suppose that the Hilbert space H is separable and $\{e_n, f_n\}_{n \in \mathbb{N}}$ is an orthonormal basis of H. Consider the closed subspace X spanned by all e_n and the closed subspace Y spanned by all $f_n + n e_n$. Every basis vector belongs to $X + Y$, which means that $X + Y$ coincides with H. The reader can show that the vector defined by a series $\sum_{n=1}^{\infty} a_n f_n$ belongs to $X + Y$ only in the case when $|n a_n|^2 \to 0$. This means that $X + Y$ is a proper subspace of H. The sum of two closed subspaces is closed if, for example, one of the subspaces is finite-dimensional or the subspaces are orthogonal.

For every subset $X \subset H$ the *orthogonal complement* $X^\perp$ is the subspace formed by all vectors orthogonal to every vector from X (this subspace is closed). If X is a closed subspace of H, then $H = X \oplus X^\perp$ and the dimension of $X^\perp$ is called the *codimension* of X. If X is a subset of H and $\overline{X}$ is the smallest closed subspace of H containing X, then $X^\perp = \overline{X}^\perp$ and $H = \overline{X} \oplus X^\perp$, which implies that $\overline{X} = X^{\perp\perp}$. In particular, for every closed subspace X we have $X^{\perp\perp} = X$. Also,

$$(X + Y)^\perp = X^\perp \cap Y^\perp \quad \text{and} \quad (X \cap Y)^\perp = X^\perp + Y^\perp$$

for any closed subspaces $X, Y \subset H$. Therefore, the orthocomplementation $X \rightarrow X^\perp$ is an involutory anti-automorphism of the lattice $\mathcal{L}(H)$. In other words, $\mathcal{L}(H)$ together with the orthocomplementation $X \rightarrow X^\perp$ is an orthomodular lattice. The remarkable Kakutani–Mackey theorem [31] states that every orthomodular lattice formed by all closed subspaces of an infinite-dimensional complex Banach space is the orthomodular lattice associated to a complex Hilbert space (see Section 3.3).

Remark 1.13 Since the lattice $\mathcal{L}(H)$ is complete and the orthocomplementation is an anti-automorphism of this lattice, for every subset $\mathcal{X} \subset \mathcal{L}(H)$ the intersection $\bigcap_{X \in \mathcal{X}} X^\perp$ coincides with the orthogonal complement of the smallest closed subspace containing all elements of $\mathcal{X}$. Similarly, the orthogonal complement of $\bigcap_{X \in \mathcal{X}} X$ is the smallest closed subspace containing all $X^\perp$ such that $X \in \mathcal{X}$.

By the definition, two closed subspaces $X, Y \subset H$ (elements of the orthomodular lattice $\mathcal{L}(H)$) are *compatible* if the subspaces

$$(X \cap Y)^\perp \cap X \quad \text{and} \quad (X \cap Y)^\perp \cap Y$$

are orthogonal. This is equivalent to the fact that there exists an orthonormal basis of H such that X and Y both are spanned by subsets of this basis. For example, $X, Y \in \mathcal{L}(H)$ are compatible if $X \subset Y$ or $X \perp Y$. Also, if $X \in \mathcal{L}(H)$ is compatible to every element of $\mathcal{L}(H)$, then X is 0 or H.

Lemma 1.14 *The following assertions are fulfilled:*

(1) *If $X, Y \in \mathcal{L}(H)$ are compatible, then X and $Y^\perp$ are compatible.*
(2) *If $X \in \mathcal{L}(H)$ is compatible to all elements of a subset $\mathcal{Y} \subset \mathcal{L}(H)$, then X is compatible to $\bigcap_{Y \in \mathcal{Y}} Y$ and the smallest closed subspace containing all elements of $\mathcal{Y}$.*

Proof (1) Consider an orthonormal basis of H such that X and Y are spanned by subsets of this basis. It is clear that $Y^\perp$ also is spanned by a subset of this basis.

(2) Suppose that $\mathcal{Y} = \{Y_1, Y_2\}$. Let $Z = Y_1 + Y_2$. For every $i = 1, 2$ we have

$$(X \cap Z)^\perp \cap X \subset (X \cap Y_i)^\perp \cap X \subset Y_i^\perp$$

(the second inclusion is equivalent to the fact that X and Y_i are compatible). Then

$$(X \cap Z)^\perp \cap X \subset Y_1^\perp \cap Y_2^\perp = Z^\perp,$$

which implies that X and Z are compatible. By the statement (1), X is compatible to both $Y_i^\perp$. Then the above arguments show that X is compatible to $Y_1^\perp + Y_2^\perp$ which coincides with $(Y_1 \cap Y_2)^\perp$. The latter means that X is compatible to $Y_1 \cap Y_2$. In the general case, the proof is similar by Remark 1.13.   □

For every orthonormal basis B of H the set of all closed subspaces spanned by subsets of B is said to be the *orthogonal apartment* associated to B. Two orthonormal bases define the same orthogonal apartment if and only if the vectors from one basis are proportional to the vectors from the other. If $\mathcal{A}$ is an orthogonal apartment of $\mathcal{L}(H)$, then the partially ordered set $(\mathcal{A}, \subset)$ is a complete Boolean algebra isomorphic to the Boolean algebra formed by all subsets of a set whose cardinality is the dimension of H.

A subset of $\mathcal{L}(H)$ will be called *compatible* if any two distinct elements from this subset are compatible.

Proposition 1.15 *If X is a compatible subset of $\mathcal{L}(H)$ formed by finite-dimensional subspaces, then there is an orthogonal apartment containing X.*

Proof First, we consider the case when H is finite-dimensional. We say that a subset $\mathcal{Y} \subset \mathcal{L}(H)$ is *special* if it consists of mutually orthogonal elements whose sum coincides with H and every element of $\mathcal{Y}$ is compatible to all elements of X. For example, for every $X \in X$ the subset $\{X, X^\perp\}$ is special by Lemma 1.14.

Let $\mathcal{Y}$ be a special subset. Suppose that there is $X \in X$ which intersects a certain $Y \in \mathcal{Y}$ in a proper subspace of Y. Then

$$Y' = Y \cap X \quad \text{and} \quad Y'' = (Y \cap X)^\perp \cap Y$$

are orthogonal subspaces whose sum coincides with Y. Lemma 1.14 shows that these subspaces are compatible to all elements of X and

$$(\mathcal{Y} \setminus \{Y\}) \cup \{Y', Y''\}$$

is a special subset. Since H is finite-dimensional, we can construct recursively a special subset $\mathcal{Z}$ such that for all $X \in \mathcal{X}$ and $Z \in \mathcal{Z}$ we have $X \cap Z = Z$ or $X \cap Z = 0$. Then every $X \in \mathcal{X}$ is the sum of all $Z \in \mathcal{Z}$ contained in it. We take any orthonormal basis of H such that every element of $\mathcal{Z}$ is spanned by a subset of this basis. The associated orthogonal apartment is as required.

Now, we assume that H is infinite-dimensional. Consider the family of all subsets of $\mathcal{L}(H)$ formed by mutually orthogonal finite-dimensional subspaces compatible to all elements of $\mathcal{X}$. Using Zorn's lemma, we establish the existence of maximal subsets with respect to this property. Such subsets will be called *special*. We assert that the closed subspace spanned by all elements of a special subset coincides with H.

Let H' be the closed subspace spanned by all elements of a special subset $\mathcal{Z}$. By Lemma 1.14, it is compatible to all elements of $\mathcal{X}$. Suppose that H' is a proper subspace of H. If all elements of $\mathcal{X}$ are contained in H', then we add to $\mathcal{Z}$ any non-zero finite-dimensional subspace of $H'^\perp$ and get a subset of $\mathcal{L}(H)$ whose elements are mutually orthogonal and compatible to all elements of $\mathcal{X}$. This is impossible, since $\mathcal{Z}$ is maximal with respect to this property. So, there is $X \in \mathcal{X}$ which is not contained in H'. Lemma 1.14 implies that the non-zero subspace

$$Y = (X \cap H')^\perp \cap X$$

is compatible to all elements of $\mathcal{X}$ and orthogonal to all elements of $\mathcal{Z}$. Then all elements of the set $\mathcal{Z} \cup \{Y\}$ are mutually orthogonal and compatible to every element of $\mathcal{X}$. We get a contradiction again. Therefore, H' coincides with H.

We take any special subset $\mathcal{Z}$ and for every $Z \in \mathcal{Z}$ denote by $\mathcal{X}_Z$ the set formed by all intersections of Z with elements of $\mathcal{X}$. This set is compatible (possibly $\mathcal{X}_Z \subset \{0, Z\}$). Since every $Z \in \mathcal{Z}$ is finite-dimensional, there is an orthonormal basis of Z such that every element of $\mathcal{X}_Z$ is spanned by a subset of this basis (if $\mathcal{X}_Z \subset \{0, Z\}$, then we can take any orthonormal basis of Z). The union of all such bases is an orthonormal basis B of H. For every $X \in \mathcal{X}$ there are finitely many $Z \in \mathcal{Z}$ such that $X \cap Z$ is non-zero. Since X is compatible to all elements of $\mathcal{Z}$, the sum of all non-zero $X \cap Z$ coincides with X. Therefore, X is contained in the orthogonal apartment associated to B. $\square$

Remark 1.16 Proposition 1.15 shows that every compatible subset of $\mathcal{L}(H)$ is contained in an orthogonal apartment if H is finite-dimensional. Is this true for the general case? We leave this question as an open problem.

The *Grassmannians* of the Hilbert space H can be defined as the orbits of the action of the group of all invertible bounded linear operators on $\mathcal{L}(H)$. Every finite-dimensional subspace of H is closed and, as above, we write $\mathcal{G}_k(H)$ for the Grassmannian formed by k-dimensional subspaces of H. We denote by $\mathcal{G}^k(H)$ the Grassmannian consisting of closed subspaces whose codimension is k. If H is infinite-dimensional, then $\mathcal{G}_\infty(H)$ denotes the set of all closed subspaces whose dimension and codimension both are infinite (this is one of the Grassmannians only in the case when H is separable). The orthocomplementation transfers every $\mathcal{G}_k(H)$ to $\mathcal{G}^k(H)$ and conversely; also, it sends every orthogonal apartment of $\mathcal{L}(H)$ to itself. The intersections of a Grassmannian with orthogonal apartments of $\mathcal{L}(H)$ are called *orthogonal apartments* of this Grassmannian. By Proposition 1.15, for each natural $k < \dim H$ every compatible subset of $\mathcal{G}_k(H)$ is contained in an orthogonal apartment. Using the orthocomplementation, we show that the same holds for compatible subsets of $\mathcal{G}^k(H)$.

A linear operator $A : H \to H$ is *bounded* if there is a non-negative real number a such that

$$\|A(x)\| \le a\|x\|$$

for all vectors $x \in N$. The smallest number a satisfying this condition is called the *norm* of A and denoted by $\|A\|$. All bounded linear operators on H form the algebra $\mathcal{B}(H)$. If A is a bounded linear operator on H, then for every vector $y \in H$ the map

$$x \to \langle A(x), y \rangle$$

is a bounded linear functional on H and, by Riesz's representation theorem, there exists the unique vector $A^*(y) \in H$ such that

$$\langle A(x), y \rangle = \langle x, A^*(y) \rangle$$

for all vectors $x \in H$. The map $A^* : H \to H$ is a bounded linear operator and $\|A^*\| = \|A\|$. This operator is known as *adjoint* to A. For any $A, B \in \mathcal{B}(H)$ we have

$$(AB)^* = B^* A^* \quad \text{and} \quad (aA + bB)^* = \bar{a}A^* + \bar{b}B^*,$$

where $a, b \in \mathbb{C}$.

A bounded linear operator A on H is called *self-adjoint* if $A^* = A$. A linear combination $aA + bB$ of self-adjoint operators A, B is self-adjoint if and only if a, b are real numbers, i.e. all self-adjoint operators form a real vector space.

Let P be an idempotent in the algebra $\mathcal{B}(H)$, i.e. $P^2 = P$. Then the restriction of P to the image $\mathrm{Im}(P)$ is identity and $x - P(x)$ belongs to the kernel $\mathrm{Ker}(P)$ for every $x \in H$. This means that H is the direct sum of $\mathrm{Ker}(P)$ and $\mathrm{Im}(P)$ (these subspaces are closed, since P is bounded), i.e. every vector $x \in H$ can be uniquely presented as the sum of two vectors $y \in \mathrm{Ker}(P)$, $z \in \mathrm{Im}(P)$ and

$P(x) = z$. The operator $\mathrm{Id}_H - P$ also is an idempotent of $\mathcal{B}(H)$ and

$$\mathrm{Ker}(\mathrm{Id}_H - P) = \mathrm{Im}(P), \quad \mathrm{Im}(\mathrm{Id}_H - P) = \mathrm{Ker}(P).$$

Note that for any pair of closed subspaces X and Y satisfying $X \oplus Y = H$ there is the unique idempotent of $\mathcal{B}(H)$ whose kernel and whose image coincide with X and Y, respectively. An easy verification shows that P is self-adjoint if and only if $\mathrm{Ker}(P)$ and $\mathrm{Im}(P)$ are orthogonal. In this case, we say that P is a *projection*. So, projections in H can be characterized as self-adjoint idempotents of the algebra $\mathcal{B}(H)$.

Example 1.17 Consider the Hilbert space of complex-valued L^2-functions on $\mathbb{R}$. This Hilbert space can be presented as the orthogonal sum of the subspace of even functions and the subspace of odd functions. Then

$$f \to \frac{f(x) + f(-x)}{2} \quad \text{and} \quad f \to \frac{f(x) - f(-x)}{2}$$

are the corresponding projections.

Every projection can be identified with its image. We denote by P_X the projection on a closed subspace X and write $\mathcal{P}_k(H)$ for the set formed by all projections of rank k, i.e. all P_X such that $X \in \mathcal{G}_k(H)$.

Proposition 1.18 *Two closed subspaces of H are orthogonal if and only if the composition of the corresponding projections is zero. Two closed subspaces of H are compatible if and only if the corresponding projections commute.*

Proof Two closed subspaces X and Y are orthogonal if and only if $P_X P_Y = 0$. In the case when these subspaces are compatible, we have

$$P_X P_Y = P_{X \cap Y} = P_Y P_X.$$

Suppose that $P_X P_Y = P_Y P_X$. Then $P = P_X P_Y$ is a projection. Indeed, we have

$$P^2 = P_X P_Y P_X P_Y = P_X P_X P_Y P_Y = P_X P_Y = P$$

and

$$P^* = (P_X P_Y)^* = P_Y^* P_X^* = P_Y P_X = P.$$

Similarly,

$$Q_1 = P_X - P \quad \text{and} \quad Q_2 = P_Y - P$$

are projections (a direct verification shows that each Q_i is an idempotent and it is self-adjoint as a real linear combination of self-adjoint operators). Let X', Y' and Z be the images of Q_1, Q_2 and P, respectively. Since

$$Q_1 Q_2 = P Q_1 = P Q_2 = 0,$$

the closed subspaces X', Y', Z are mutually orthogonal. Then

$$P_X = P + Q_1 = P_Z + P_{X'} = P_{Z+X'},$$

which implies that $X = Z + X'$. Similarly, we establish that $Y = Z + Y'$. Therefore, X and Y are compatible. □

Remark 1.19 By the general spectral theorem, every bounded self-adjoint operator on H can be identified with a unique spectral measure which takes values in the set of projections or, equivalently, in the lattice $\mathcal{L}(H)$. Two spectral measures are called *compatible* if all values of one measure are compatible to all values of the other. Then two bounded self-adjoint operators commute if and only if the corresponding measures are compatible [63, Theorem 4.11].

It follows from the spectral theorem (the finite-dimensional version) that every self-adjoint finite-rank operator on H can be presented as a linear combination of rank one projections whose images are mutually orthogonal, and all scalars in this linear combination are real numbers.

Proposition 1.20 *Let k be a positive integer less than* $\dim H$*. Then every self-adjoint finite-rank operator on H can be presented as a linear combination of rank-k projections whose images are mutually compatible, i.e. these projections mutually commute, and all scalars in this linear combination are real numbers.*

Proof For every self-adjoint finite-rank operator there is an orthonormal basis B of H such that this operator is a real linear combination of the projections on some 1-dimensional subspaces containing vectors from B. Let $e_1, \ldots, e_{k+1}$ be vectors from B. Denote by P_i the projection on the k-dimensional subspace spanned by all e_j with $j \neq i$. Then

$$\frac{1}{k}[P_2 + \cdots + P_{k+1} - (k-1)P_1]$$

is the projection on the 1-dimensional subspace containing e_1. Similarly, we show that every projection on the 1-dimensional subspace containing a vector from B is a real linear combination of the projections on some k-dimensional subspaces spanned by subsets of B. □

Let A be a bounded linear operator on H such that for every orthonormal basis $\{e_i\}_{i \in I}$ of H the set $\{i \in I : A(e_i) \neq 0\}$ is countable. This operator is said to be of *trace class* if for every orthonormal basis $\{e_i\}_{i \in I}$ of H we have

$$\sum_{i \in I} |\langle A(e_i), e_i \rangle| < \infty.$$

The fact that this condition holds for one orthonormal basis does not imply that the same holds for all orthonormal bases.

Example 1.21 Suppose that H is separable and $\{e_i\}_{i\in\mathbb{N}}$ is an orthonormal basis of H. If A is the linear operator on H such that $A(e_i) = e_{i+1}$ for every $i \in \mathbb{N}$, then $\langle A(e_i), e_i\rangle = 0$ for all i and the corresponding series is zero. For every $j \in \mathbb{N}$ we take

$$f_{2j-1} = \frac{e_{2j-1} + e_{2j}}{\sqrt{2}}, \quad f_{2j} = \frac{e_{2j-1} - e_{2j}}{\sqrt{2}}$$

and consider the orthonormal basis $\{f_i\}_{i\in\mathbb{N}}$. We have

$$A(f_{2j-1}) = \frac{e_{2j} + e_{2j+1}}{\sqrt{2}}, \quad A(f_{2j}) = \frac{e_{2j} - e_{2j+1}}{\sqrt{2}}$$

and

$$\langle A(f_{2j-1}), f_{2j-1}\rangle = 1/2, \quad \langle A(f_{2j}), f_{2j}\rangle = -1/2.$$

The series $\sum_{i\in\mathbb{N}} |\langle A(f_i), f_i\rangle|$ does not converge.

If A is an operator of trace class, then the sum

$$\text{tr}(A) = \sum_{i\in I}{}' \langle A(e_i), e_i\rangle$$

does not depend on an orthonormal basis $\{e_i\}_{i\in I}$ and is called the *trace* of the operator A. In this case, the adjoint operator A^* also is of trace class and $\text{tr}(A) = \overline{\text{tr}(A^*)}$. If A and B are operators of trace class, then for any scalars $a, b \in \mathbb{C}$ the operator $aA + bB$ is of trace class and

$$\text{tr}(aA + bB) = a \cdot \text{tr}(A) + b \cdot \text{tr}(B),$$

i.e. the trace function is linear. If $A, B \in \mathcal{B}(H)$ and A is of trace class, then AB and BA both are of trace class and $\text{tr}(AB) = \text{tr}(BA)$, i.e. all operators of trace class form an ideal in the algebra $\mathcal{B}(H)$.

A bounded linear operator A on H is called *positive* if $\langle A(x), x\rangle \geq 0$ for all $x \in H$. In this case, A is self-adjoint and there is a unique positive operator B on H such that $A = B^2$ [55, Theorems 12.32 and 12.33].

Denote by $\mathcal{J}(H)$ the set of all operators $A \in \mathcal{B}(H)$ satisfying the following conditions:

- A is positive,
- A is of trace class and $\text{tr}(A) = 1$.

Since we have

$$tA + (1 - t)B \in \mathcal{J}(H) \quad \text{for all} \quad A, B \in \mathcal{J}(H) \quad \text{and} \quad t \in [0, 1],$$

the set $\mathcal{J}(H)$ is convex. The trace of a rank-k projection is equal to k and $\mathcal{P}_k(H)$ is contained in $\mathcal{J}(H)$ only in the case when $k = 1$. Every $A \in \mathcal{J}(H)$ can be presented as the sum

$$A = \sum_{i \in I} t_i P_i, \tag{1.2}$$

where I is a countable set, $\{t_i\}_{i \in I}$ are positive real numbers satisfying $\sum_{i \in I} t_i = 1$ and $\{P_i\}_{i \in I}$ are projections on mutually orthogonal 1-dimensional subspaces.

Proposition 1.22 $\mathcal{P}_1(H)$ *is the set of all extreme points of the convex set* $\mathcal{J}(H)$.

Proof   Let P be the projection on a 1-dimensional subspace $X \subset H$. Suppose that

$$P = t A_1 + (1 - t) A_2$$

for some $A_1, A_2 \in \mathcal{J}(H)$ and $t \in (0, 1)$. Since

$$0 = \langle P(x), x \rangle = t \langle A_1(x), x \rangle + (1 - t) \langle A_2(x), x \rangle$$

for all $x \in X^\perp$ and A_1, A_2 are positive, we have $\langle A_i(x), x \rangle = 0$ for every $x \in X^\perp$ and $i = 1, 2$. If B_i is the positive operator satisfying $A_i = B_i^2$, then

$$0 = \langle A_i(x), x \rangle = \langle B_i^2(x), x \rangle = \langle B_i(x), B_i(x) \rangle$$

and $B_i(x) = 0$ for all $x \in X^\perp$. This means that $A_i(x) = 0$ for all $x \in X^\perp$; in other words, A_i is a scalar multiple of P. Since $\mathrm{tr}(A_i) = 1$, we obtain that $P = A_1 = A_2$, which implies that P is an extreme point of $\mathcal{J}(H)$.

Consider $A \in \mathcal{J}(H)$ as the sum (1.2) and suppose that $|I| \geq 2$, i.e. A is not the projection on a 1-dimensional subspace. For every $i \in I$ we have

$$A = t_i P_i + (1 - t_i) B,$$

where

$$B = \sum_{j \in I \setminus \{i\}} \frac{t_j}{1 - t_i} P_j$$

belongs to $\mathcal{J}(H)$, i.e. A is not an extreme point of $\mathcal{J}(H)$. $\square$

For every operator $A \in \mathcal{J}(H)$ we define the function $p_A : \mathcal{L}(H) \to [0, 1]$ as

$$p_A(X) = \mathrm{tr}(A P_X)$$

for all $X \in \mathcal{L}(H)$ (note that $\mathrm{tr}(A P_X) = \mathrm{tr}(P_X A)$). This is a state on the ortho-modular lattice $\mathcal{L}(H)$.

Theorem 1.23 (Gleason [26])   *If H is separable and* $\dim H \geq 3$, *then for every state p on $\mathcal{L}(H)$ there is $A \in \mathcal{J}(H)$ such that $p = p_A$.*

For any $A, B \in \mathcal{J}(H)$ and $t \in [0, 1]$ we have

$$p_{tA+(1-t)B} = t p_A + (1 - t) p_B .$$

Therefore, Gleason's theorem shows that the convex set of all states of the orthomodular lattice $\mathcal{L}(H)$ can be identified with the convex set $\mathcal{J}(H)$ and pure states correspond to rank one projections if H is a separable complex Hilbert space of dimension not less than three.

For the case when $\dim H = 2$ the above statement fails.

Example 1.24   Let $\dim H = 2$. Consider two disjoint subsets $\mathcal{X}_1, \mathcal{X}_2 \subset \mathcal{G}_1(H)$ such that $\mathcal{X}_1 \cup \mathcal{X}_2 = \mathcal{G}_1(H)$ and for every $X \in \mathcal{X}_i$ the orthogonal complement $X^\perp$ belongs to $\mathcal{X}_{3-i}$. The function $p : \mathcal{L}(H) \to [0, 1]$ defined as

$$p(X) = \begin{cases} 0 & X = 0 \text{ or } X \in \mathcal{X}_1, \\ 1 & X = H \text{ or } X \in \mathcal{X}_2 \end{cases}$$

is a state which cannot be obtained from $A \in \mathcal{J}(H)$.

In the case when H is non-separable, the statement of Gleason's theorem holds if and only if the dimension of H is a non-measurable cardinality [19, Section 3.5].

2

Geometric Transformations of Grassmannians

In this chapter, we present some geometrical characterizations of semilinear isomorphisms of vector spaces. It must be pointed out that we are interested only in results which hold for vector spaces of an arbitrary (not necessarily finite) dimension.

We begin from the classic version of the Fundamental Theorem of Projective Geometry which states that all isomorphisms between the lattices formed by subspaces of vector spaces are induced by semilinear isomorphisms of these vector spaces (see, for example, [3]) or, equivalently, every isomorphism of the projective spaces associated to vector spaces is induced by a semilinear isomorphism between these vector spaces. A modern version of the Fundamental Theorem of Projective Geometry [21, 28] describes a more general class of collinearity preserving maps of projective spaces in terms of semilinear maps associated to not necessarily surjective homomorphisms of division rings.

There are analogues of the Fundamental Theorem of Projective Geometry for Grassmannians formed by subspaces of dimension and codimension greater than one. The first result of such kind is the classic Chow's theorem [13] concerning automorphisms of the Grassmann graphs of finite-dimensional vector spaces; it states that every such automorphism can be uniquely extended to an automorphism or an anti-automorphism of the lattice of subspaces; in other words, it is induced by a semilinear automorphism of the vector space or a semilinear isomorphism of the vector space to the dual vector space.

For an infinite-dimensional vector space there are the following two types of Grassmannians:

(1) Grassmannians formed by subspaces of finite dimension or codimension,
(2) Grassmannians consisting of subspaces whose dimension and codimension both are infinite.

Grassmannians of the first type are similar to Grassmannians of finite-dimensional vector spaces, and all automorphisms of the corresponding Grassmann graphs are induced by semilinear automorphisms of the vector space or semilinear automorphisms of the dual vector space. The Grassmann graphs associated to Grassmannians of the second type are not connected and admit automorphisms which are not induced by semilinear automorphisms of the vector space. For this reason, the problem is formulated as follows: describe all restrictions of automorphisms of Grassmann graphs to connected components. This problem is still open.

Consider the Grassmannian formed by subspaces whose dimension is equal to the codimension. Note that complements of such subspaces also belong to this Grassmannian. By [8], every bijective transformation preserving the complementarity relation in both directions is an automorphism of the associated Grassmann graph. If the vector space is finite-dimensional, then its dimension is even and the Grassmannian consists of subspaces whose dimension is equal to half of the dimension. In this case, every complementary preserving transformation can be uniquely extended to an automorphism or an anti-automorphism of the lattice of subspaces.

Another geometrical characterization of semilinear automorphisms is related to the concept of apartments in Grassmannians. The transformations of Grassmannians induced by semilinear automorphisms can be characterized as bijections which preserve apartments in both directions [17].

2.1 Semilinear Maps

Let V and V' be left vector spaces over division rings R and R', respectively. We suppose that the dimensions of these vector spaces are not less than three.

A map $L : V \to V'$ is called *semilinear* if

$$L(x + y) = L(x) + L(y)$$

for all vectors $x, y \in V$ and there is a non-zero homomorphism $\sigma : R \to R'$ such that

$$L(ax) = \sigma(a)L(x)$$

for all vectors $x \in V$ and all scalars $a \in R$. In this case, we will say also that L is *σ-linear*. It is easy to see that every non-zero homomorphism between division rings is injective. We refer to [45, Chapter 1] for examples and basic properties of semilinear maps.

For a semilinear map $L : V \to V'$ every scalar multiple aL also is semilinear.

The corresponding homomorphism of division rings is $b \rightarrow a\sigma(b)a^{-1}$, where σ is the homomorphism associated to L.

A semilinear map is said to be a *semilinear isomorphism* if it is bijective and the associated homomorphism of division rings is an isomorphism. Every semilinear isomorphism $L : V \rightarrow V'$ preserves the linear dependence in both directions and we have $\dim V = \dim V'$. Therefore, L induces an isomorphism between the lattices $\mathcal{L}(V)$ and $\mathcal{L}(V')$. Any non-zero scalar multiple of L induces the same lattice isomorphism.

Example 2.1   There are semilinear bijections which are not semilinear isomorphisms. Consider, for example, the semilinear bijection of $\mathbb{R}^2$ to $\mathbb{C}$ sending every vector (a, b) to the complex number $a + bi$. The associated homeomorphism of division rings is the natural embedding of the real field in the field of complex numbers. Another example is the semilinear bijection of $\mathbb{R}^4$ to $\mathbb{H}$ transferring every vector (a, b, c, d) to the quaternion $a + b\mathbf{i} + c\mathbf{j} + d\mathbf{k}$.

Lemma 2.2   *Let $\mathcal{G}$ be one of the Grassmannians of V. If L and S are semilinear isomorphisms of V to V' such that $L(X) = S(X)$ for every $X \in \mathcal{G}$, then one of these semilinear isomorphisms is a scalar multiple of the other.*

Proof   Since every $P \in \mathcal{G}_1(V)$ is the intersection of all elements from $\mathcal{G}$ containing P, we have $L(P) = S(P)$ for all $P \in \mathcal{G}_1(V)$. The latter implies that for every non-zero vector $x \in V$ there is a non-zero scalar a_x such that

$$S(x) = a_x L(x).$$

If $x, y \in V$ are linearly independent, then

$$a_x L(x) + a_y L(y) = S(x + y) = a_{x+y}(L(x) + L(y)),$$

which means that $a_x = a_{x+y} = a_y$, since $L(x)$ and $L(y)$ are linearly independent. If y is a scalar multiple of x, then we take any vector $z \in V$ such that x, z are linearly independent and establish that $a_x = a_z = a_y$. Therefore, $a_x = a_y$ for any two non-zero vectors $x, y \in V$ and we get the claim.   □

Theorem 2.3 (Fundamental Theorem of Projective Geometry)   *Every isomorphism of the lattice $\mathcal{L}(V)$ to the lattice $\mathcal{L}(V')$ is induced by a semilinear isomorphism $L : V \rightarrow V'$, and any other semilinear isomorphism inducing this lattice isomorphism is a scalar multiple of L.*

The proof of this theorem follows Lemma 2.10.

Remark 2.4   In the case when $\dim V = \dim V' = 2$, every bijection f of $\mathcal{G}_1(V)$ to $\mathcal{G}_1(V')$ (if it exists) can be uniquely extended to an isomorphism or an anti-isomorphism of $\mathcal{L}(V)$ to $\mathcal{L}(V')$ (we set $f(0) = 0, f(V) = V'$ or $f(0) = V', f(V) = 0$, respectively).

Suppose that V is finite-dimensional. If the division ring R is isomorphic to the opposite division ring R^* (in particular, if R is commutative), then there exist semilinear isomorphisms of V to V^*. Every such semilinear isomorphism induces an isomorphism between the lattices $\mathcal{L}(V)$ and $\mathcal{L}(V^*)$. The composition of this lattice isomorphism and the annihilator map is an anti-automorphism of the lattice $\mathcal{L}(V)$. The following statement is a simple consequence of Theorem 2.3.

Corollary 2.5 *If V is finite-dimensional, then every anti-automorphism of the lattice $\mathcal{L}(V)$ is induced by a semilinear isomorphism $L : V \rightarrow V^*$, i.e. it is the composition of the isomorphism of $\mathcal{L}(V)$ to $\mathcal{L}(V^*)$ induced by L and the annihilator map. Any other semilinear isomorphism inducing this anti-automorphism is a scalar multiple of L.*

A *point-line geometry* is a pair $(\mathcal{P}, \mathcal{L})$, where $\mathcal{P}$ is a non-empty set whose elements are called *points* and $\mathcal{L}$ is a family of subsets of $\mathcal{P}$ called *lines* and satisfying some axioms. Denote by Π_V the projective space associated to the vector space V (recall that the dimension of V is assumed to be not less than three). This is the point-line geometry whose points are elements of $\mathcal{G}_1(V)$ and whose lines are defined by elements of $\mathcal{G}_2(V)$, i.e. for every $S \in \mathcal{G}_2(V)$ the associated line is $\mathcal{G}_1(S)$. An *isomorphism* between point-line geometries $(\mathcal{P}, \mathcal{L})$ and $(\mathcal{P}', \mathcal{L}')$ is a bijection $f : \mathcal{P} \rightarrow \mathcal{P}'$ such that $f(\mathcal{L}) = \mathcal{L}'$.

Let $L : V \rightarrow V'$ be a semilinear injection. For every 1-dimensional subspace $P \subset V$ the image $L(P)$ is a non-zero subset in a certain 1-dimensional subspace of V'. The set $L(P)$ is a subspace of V' only in the case when the division ring homomorphism associated to L is an isomorphism, i.e. L is a semilinear isomorphism to a subspace of V'. Consider the map

$$(L)_1 : \mathcal{G}_1(V) \rightarrow \mathcal{G}_1(V')$$

which sends every $P \in \mathcal{G}_1(V)$ to the element of $\mathcal{G}_1(V')$ containing $L(P)$.

This map is not necessarily injective. Indeed, if for some distinct 1-dimensional subspaces $P, Q \subset V$ the images $L(P)$ and $L(Q)$ are contained in the same 1-dimensional subspace of V', then the restriction of $(L)_1$ to the line $\mathcal{G}_1(P + Q)$ is constant. In particular, if this holds for any two 1-dimensional subspaces $P, Q \subset V$, then $(L)_1$ is constant. In the case when $(L)_1$ transfers P and Q to distinct elements of $\mathcal{G}_1(V')$, the restriction of $(L)_1$ to the line $\mathcal{G}_1(P + Q)$ is injective.

For every non-zero scalar $a \in R'$ we have $(aL)_1 = (L)_1$. Conversely, if $(L)_1$ is non-constant and $(L)_1 = (S)_1$ for a semilinear injection $S : V \to V'$, then S is a non-zero scalar multiple of L (the proof is similar to the proof of Lemma 2.2).

The following version of the Fundamental Theorem of Projective Geometry was proved by Faure and Frölicher [21] and Havlicek [28] independently.

Theorem 2.6 (Faure and Frölicher [21] and Havlicek [28])   *Suppose that a map* $f : \mathcal{G}_1(V) \to \mathcal{G}_1(V')$ *satisfies the following conditions:*

(1) *for mutually distinct* $P, P_1, P_2 \in \mathcal{G}_1(V)$ *the inclusion* $P \subset P_1 + P_2$ *implies that* $f(P) \subset f(P_1) + f(P_2)$;
(2) *the image of* f *is not contained in a line of* $\Pi_{V'}$.

Then f *is induced by a semilinear injection* $L : V \to V'$ *and any other semilinear map inducing* f *is a scalar multiple of* L.

We will use the following simple modification of Theorem 2.6.

Corollary 2.7   *Suppose that* $f : \mathcal{G}_1(V) \to \mathcal{G}_1(V')$ *is an injection which transfers lines to subsets of lines and satisfies the condition* (2) *from Theorem 2.6. Then* f *is induced by a semilinear injection* $L : V \to V'$ *and any other semilinear map inducing* f *is a scalar multiple of* L.

Remark 2.8   It is clear that every injection $f : \mathcal{G}_1(V) \to \mathcal{G}_1(V')$ transferring lines to subsets of lines satisfies the condition (1) from Theorem 2.6. However, there are maps of $\mathcal{G}_1(V)$ to $\mathcal{G}_1(V')$ sending lines to subsets of lines and whose restrictions to some lines are non-injective and non-constant [45, Example 2.3]. Such maps cannot be induced by semilinear maps.

Corollary 2.9   *Every isomorphism* f *between the projective spaces* Π_V *and* $\Pi_{V'}$ *is induced by a semilinear isomorphism* $L : V \to V'$ *and any other semilinear map inducing* f *is a scalar multiple of* L.

Corollary 2.9 is a simple consequence of Corollary 2.7 and the following.

Lemma 2.10   *If* $L : V \to V'$ *is a semilinear injection such that* $(L)_1$ *is bijective and transfers lines of* Π_V *to lines of* $\Pi_{V'}$, *then* L *is a semilinear isomorphism.*

Proof   The restriction of $(L)_1$ to every line $\mathcal{G}_1(S)$, $S \in \mathcal{G}_2(V)$ is a bijection to a line $\mathcal{G}_1(S')$, $S' \in \mathcal{G}_2(V')$. This implies that the division ring homomorphism associated to L is an isomorphism. Then $L(V)$ is a subspace of V'. Since $(L)_1$ is bijective, we have $L(V) = V'$.   □

Proof of Theorem 2.3   Let f be an isomorphism between the lattices $\mathcal{L}(V)$ and $\mathcal{L}(V')$. The restriction of f to $\mathcal{G}_1(V)$ is an isomorphism of Π_V to $\Pi_{V'}$. By

Corollary 2.9, it is induced by a semilinear isomorphism $L : V \to V'$ (unique up to a scalar multiple), i.e. $f(P) = L(P)$ for every $P \in \mathcal{G}_1(V)$. Since $P \in \mathcal{G}_1(V)$ is contained in $X \in \mathcal{L}(V)$ if and only if $f(P) = L(P)$ is contained in $f(X)$, we have $f(X) = L(X)$ for all $X \in \mathcal{L}(V)$. □

The *dual* projective space Π_V^* is the point-line geometry whose points are elements of $\mathcal{G}^1(V)$ and whose lines are defined by elements of $\mathcal{G}^2(V)$, i.e. for every $S \in \mathcal{G}^2(V)$ the associated line consists of all elements of $\mathcal{G}^1(V)$ containing S. The annihilator map is an isomorphism between the projective spaces Π_V^* and Π_{V^*}.

The description of automorphisms of the dual projective space is related to the concept of adjoint map. Let L be a semilinear automorphism of V and let σ be the associated automorphism of the division ring R. The *adjoint* map L^* is the transformation of V^* defined by the equality

$$L^*(x^*)[x] = \sigma^{-1}(x^*[L(x)])$$

for all vectors $x \in V$ and $x^* \in V^*$. This is a (σ^{-1})-linear automorphism of V^* and

$$(L^{-1})^* = (L^*)^{-1}.$$

If S is a semilinear automorphism of V, then

$$(LS)^* = S^* L^*.$$

This implies that the *contragradient* map $L \to (L^{-1})^*$ is a monomorphism of the group formed by semilinear automorphisms of V to the group of semilinear automorphisms of V^*. The contragradient map is a group isomorphism only in the case when V is finite-dimensional. If V is infinite-dimensional, then there exist semilinear automorphisms of V^* which are not adjoint to semilinear automorphisms of V. Indeed, every adjoint map preserves the family of hyperplanes P^0, $P \in \mathcal{G}_1(V)$. In the case when V is infinite-dimensional, V^* contains hyperplanes which do not belong to this family. Every semilinear automorphism of V^* transferring such a hyperplane to a hyperplane P^0 cannot be adjoint to a semilinear isomorphism of V.

If f is an automorphism of the dual projective space Π_V^*, then the transformation

$$Y \to f(^0 Y)^0, \quad Y \in \mathcal{G}_1(V^*)$$

is an automorphism of the projective space Π_{V^*}. It is induced by a semilinear automorphism of V^*. Therefore, every automorphism of Π_V^* is a transformation of type

$$X \to {}^0 S(X^0), \quad X \in \mathcal{G}^1(V),$$

where S is a semilinear automorphism of V^*. If S is the contragradient of a certain semilinear automorphism L of V, i.e. $S = (L^{-1})^*$, then

$$^0S(X^0) = L(X)$$

for every subspace $X \subset V$ (we leave all details for readers). In particular, every automorphism of Π_V^* is induced by a semilinear automorphism of V in the case when V is finite-dimensional. If V is infinite-dimensional, then there are automorphisms of Π_V^* which are not induced by semilinear automorphisms of V.

2.2 Proof of Theorem 2.6

As in the previous section, we suppose that V and V' are left vector spaces over division rings R and R', respectively. The dimensions of these vector spaces are assumed to be not less than three. For every non-zero vector x in V or V' we denote by $\langle x \rangle$ the 1-dimensional subspace containing this vector.

Let $f : \mathcal{G}_1(V) \to \mathcal{G}_1(V')$ be a map satisfying the conditions of Theorem 2.6, i.e. for any mutually distinct $P, P_1, P_2 \in \mathcal{G}_1(V)$ we have

$$P \subset P_1 + P_2 \implies f(P) \subset f(P_1) + f(P_2)$$

and the image of f is not contained in a line of $\Pi_{V'}$. The first condition implies that the restriction of f to every line $\mathcal{G}_1(S)$, $S \in \mathcal{G}_2(V)$ is injective or constant. We show that f is induced by a semilinear injection $L : V \to V'$ in several steps.

1. By our assumption, there are $P_1, P_2, P_3 \in \mathcal{G}_1(V)$ such that

$$Q_1 = f(P_1), \quad Q_2 = f(P_3), \quad Q_3 = f(P_3)$$

are not contained in a line of $\Pi_{V'}$. Let $x_1 \in P_1, x_2 \in P_2, x_3 \in P_3$ be non-zero vectors. We find non-zero vectors $y_1 \in Q_1, y_2 \in Q_2, y_3 \in Q_3$ such that

$$f(\langle x_i + x_j \rangle) = \langle y_i + y_j \rangle$$

for any $i, j \in \{1, 2, 3\}$ and

$$f(\langle x_1 + x_2 + x_3 \rangle) = \langle y_1 + y_2 + y_3 \rangle.$$

We take any non-zero $y_1 \in Q_1$. The 1-dimensional subspace $f(\langle x_1 + x_2 \rangle)$ is contained in $Q_1 + Q_2$ and the restriction of f to the line $\mathcal{G}_1(P_1 + P_2)$ is injective (since $Q_1 \neq Q_2$). This implies the existence of non-zero $y_2 \in Q_2$ satisfying

$$f(\langle x_1 + x_2 \rangle) = \langle y_1 + y_2 \rangle.$$

Similarly, we find non-zero $y_3 \in Q_3$ such that

$$f(\langle x_1 + x_3 \rangle) = \langle y_1 + y_3 \rangle.$$

Since $\langle x_1 + x_2 + x_3 \rangle$ is the intersection of

$$\langle x_1 + x_2 \rangle + P_3 \quad \text{and} \quad \langle x_1 + x_3 \rangle + P_2,$$

$f(\langle x_1 + x_2 + x_3 \rangle)$ is contained in the intersection of

$$\langle y_1 + y_2 \rangle + Q_3 \quad \text{and} \quad \langle y_1 + y_3 \rangle + Q_2$$

which coincides with $\langle y_1 + y_2 + y_3 \rangle$. Also, we have

$$\langle x_2 + x_3 \rangle = (P_2 + P_3) \cap (P_1 + \langle x_1 + x_2 + x_3 \rangle)$$

and $f(\langle x_2 + x_3 \rangle)$ is contained in the subspace

$$(Q_2 + Q_3) \cap (Q_1 + \langle y_1 + y_2 + y_3 \rangle)$$

which coincides with $\langle y_2 + y_3 \rangle$.

2. For every non-zero vector $x \in V$ we take any $i \in \{1, 2, 3\}$ such that $f(\langle x \rangle)$ is distinct from Q_i. Since $f(\langle x_i + x \rangle)$ is contained in $Q_i + f(\langle x \rangle)$ and the restriction of f to the line $\mathcal{G}_1(P_i + \langle x \rangle)$ is injective, there is a non-zero vector $L_i(x) \in f(\langle x \rangle)$ satisfying

$$f(\langle x_i + x \rangle) = \langle y_i + L_i(x) \rangle.$$

If $f(\langle x \rangle)$ is distinct from all Q_i, then $L_i(x)$ is defined for every $i \in \{1, 2, 3\}$. In the case when $f(\langle x \rangle)$ coincides with Q_k, we have only two $L_i(x)$ for $i \neq k$.

Suppose that $f(\langle x \rangle)$ is distinct from Q_i and Q_j. In the case when $f(\langle x \rangle)$ is not contained in $Q_i + Q_j$, we apply the arguments from the first step to the vectors $y_i, y_j, L_i(x)$ (instead of the vectors y_1, y_2, y_3) and establish that

$$f(\langle x_j + x \rangle) = \langle y_j + L_i(x) \rangle,$$

which means that $L_i(x) = L_j(x)$. If $f(\langle x \rangle) \subset Q_i + Q_j$, then $f(\langle x \rangle)$ is distinct from the remaining Q_k with $k \neq i, j$. Applying the same arguments to the triples $y_i, y_k, L_i(x)$ and $y_j, y_k, L_j(x)$, we obtain that $L_i(x) = L_k(x)$ and $L_j(x) = L_k(x)$.

So, if $f(\langle x \rangle)$ is distinct from all Q_i, then

$$L_1(x) = L_2(x) = L_3(x)$$

and we denote this vector by $L(x)$. In the case when $f(\langle x \rangle)$ coincides with Q_k, the vector $L_k(x)$ is not defined, but for the remaining two $i, j \in \{1, 2, 3\} \setminus \{k\}$ we have

$$L_i(x) = L_j(x)$$

and denote this vector by $L(x)$. The equality $f(\langle x_i + x_j \rangle) = \langle y_i + y_j \rangle$ shows that $L(x_i) = y_i$ for every i.

We put $L(0) = 0$ and consider the map $L : V \to V'$.

3. Show that the map L is additive, i.e. $L(x+y) = L(x)+L(y)$ for any $x, y \in V$.

Suppose that the vectors $L(x)$ and $L(y)$ are linearly independent. There exists $i \in \{1, 2, 3\}$ such that Q_i is not contained in $f(\langle x \rangle) + f(\langle y \rangle)$. Then

$$f(\langle x_i + x \rangle) = \langle y_i + L(x) \rangle \ \text{ and } \ f(\langle x_i + y \rangle) = \langle y_i + L(y) \rangle.$$

As in the first step, we establish that

$$f(\langle x + y \rangle) = \langle L(x) + L(y) \rangle \ \text{ and } \ f(\langle x_i + x + y \rangle) = \langle y_i + L(x) + L(y) \rangle,$$

in other words,

$$\langle L(x + y) \rangle = \langle L(x) + L(y) \rangle \ \text{ and } \ \langle y_i + L(x + y) \rangle = \langle y_i + L(x) + L(y) \rangle.$$

The first equality shows that $L(x + y)$ is a scalar multiple of $L(x) + L(y)$ and the second implies that the corresponding scalar is 1.

If $L(x)$ is a scalar multiple of $L(y)$, then $f(\langle x \rangle) = f(\langle y \rangle)$ coincides with $f(\langle x + y \rangle)$. We take any vector $z \in V$ such that $L(y)$ and $L(z)$ are linearly independent. Then $L(y + z) = L(y) + L(z)$ and

$$L(x + y + z) = L(x + y) + L(z), \tag{2.1}$$

since $L(x+y) \in f(\langle x \rangle) = f(\langle y \rangle)$ and $L(z)$ also are linearly independent. We have $f(\langle y \rangle) \neq f(\langle z \rangle)$, which means that the restriction of f to the line $G_2(\langle y \rangle + \langle z \rangle)$ is injective. Therefore, $f(\langle x \rangle) = f(\langle y \rangle)$ is distinct from $f(\langle y + z \rangle)$, i.e. the vectors $L(x)$ and $L(y + z)$ are linearly independent and

$$L(x + y + z) = L(x) + L(y + z) = L(x) + L(y) + L(z).$$

The required equality follows from (2.1).

4. For every non-zero scalar $a \in R$ consider the map $x \to L(ax)$. Since $L(x)$ and $L(ax)$ belong to $f(\langle x \rangle) = f(\langle ax \rangle)$, for every non-zero vector $x \in V$ there is a non-zero scalar $a_x \in R'$ such that

$$L(ax) = a_x L(x).$$

Using the fact that the map $x \to L(ax)$ is additive and its image contains linearly independent vectors, we establish that a_x does not depend on x (see the proof of Lemma 2.2 for the details) and denote this scalar by $\sigma(a)$. We put $\sigma(0) = 0$ and get a map $\sigma : R \to R'$ satisfying

$$L(ax) = \sigma(a)L(x)$$

for all $a \in R$ and $x \in V$. Then

$$\sigma(a+b)L(x) = L(ax+bx) = L(ax) + L(bx) = (\sigma(a) + \sigma(b))L(x),$$

$$\sigma(ab)L(x) = L(abx) = \sigma(a)\sigma(b)L(x)$$

and σ is a homomorphism of division rings.

So, the map L is semilinear. Since $L(x)$ is a non-zero vector belonging to $f(\langle x \rangle)$ for every non-zero $x \in V$, this semilinear map is injective and $f = (L)_1$.

2.3 Grassmann Graphs

Let V be a left vector space over a division ring. The dimension of V is assumed to be not less than three.

Let $\mathcal{G}$ be one of the Grassmannians of V, i.e. $\mathcal{G}$ coincides with $\mathcal{G}_\alpha(V)$ or $\mathcal{G}^\alpha(V)$, where α is a cardinality not greater than $\dim V$. We say that elements $X, Y \in \mathcal{G}$ are *adjacent* if

$$\dim X/(X \cap Y) = \dim Y/(X \cap Y) = 1.$$

This is equivalent to

$$\dim(X+Y)/X = \dim(X+Y)/Y = 1.$$

Note that any two distinct elements of $\mathcal{G}_1(V)$ are adjacent and the same holds for distinct elements of $\mathcal{G}^1(V)$. The associated *Grassmann graph* $\Gamma(\mathcal{G})$ is the graph whose vertex set is $\mathcal{G}$ and whose edges are pairs of adjacent elements. This graph is denoted by $\Gamma_\alpha(V)$ if $\mathcal{G} = \mathcal{G}_\alpha(V)$. In the case when $\mathcal{G} = \mathcal{G}^\alpha(V)$, we will write $\Gamma^\alpha(V)$.

Let k be a natural number less than $\dim V$ (if V is infinite-dimensional, then it is an arbitrary natural number). Two elements of $\mathcal{G}_k(V)$ are adjacent if and only if their intersection is $(k-1)$-dimensional or, equivalently, their sum is $(k+1)$-dimensional. Suppose that $\dim V = n$ is finite. For any $X, Y \in \mathcal{G}_k(V)$ we have

$$(X \cap Y)^0 = X^0 + Y^0$$

and the dimension of $X \cap Y$ is equal to the codimension of $X^0 + Y^0$. This implies that X and Y are adjacent if and only if X^0 and Y^0 are adjacent elements of $\mathcal{G}_{n-k}(V^*)$, i.e. the annihilator map is an isomorphism of $\Gamma_k(V)$ to $\Gamma_{n-k}(V^*)$.

The *distance* $d(v, w)$ between two vertices v, w in a connected graph is the smallest number i such that there is a path from v to w which consists of i edges. A path between v and w is said to be a *geodesic* if it contains precisely $d(v, w)$ edges.

Proposition 2.11   *For every natural $k < \dim V$ the Grassmann graph $\Gamma_k(V)$ is connected and the distance between $X, Y \in \mathcal{G}_k(V)$ in $\Gamma_k(V)$ is equal to*

$$k - \dim(X \cap Y) = \dim(X + Y) - k.$$

Proof   Let X and Y be distinct elements of $\mathcal{G}_k(V)$. The statement is trivial if they are adjacent and we suppose that $\dim(X \cap Y) < k - 1$. Consider a basis

$$x_1, \dots, x_m, y_1, \dots, y_m, z_1, \dots, z_{k-m}, \quad m = k - \dim(X \cap Y)$$

of $X + Y$ such that X and Y are spanned by

$$x_1, \dots, x_m, z_1, \dots, z_{k-m} \text{ and } y_1, \dots, y_m, z_1, \dots, z_{k-m},$$

respectively (in the case when $X \cap Y = 0$, this basis consists of x_i and y_i with $i \in \{1, \dots, k\}$ and does not contain z_i). For every $i \in \{1, \dots, m - 1\}$ we denote by X_i the k-dimensional subspace spanned by

$$y_1, \dots, y_i, x_{i+1}, \dots, x_m, z_1, \dots, z_{k-m}$$

and get the path $X, X_1, \dots, X_{m-1}, Y$ in $\Gamma_k(V)$. This means that the distance $d(X, Y)$ is not greater than $m = k - \dim(X \cap Y)$.

Let $X = X_0, X_1, \dots, X_i = Y$ be a geodesic in $\Gamma_k(V)$. Using induction, we show that

$$\dim(X + X_j) \le k + j$$

for every $j \in \{1, \dots, i\}$. Then

$$\dim(X + Y) \le k + i,$$

which implies that

$$d(X, Y) = i \ge \dim(X + Y) - k = k - \dim(X \cap Y).$$

So, $d(X, Y)$ is equal to $k - \dim(X \cap Y)$. $\square$

In the case when $\dim V \ge 2k$, Proposition 2.11 implies that the diameter of $\Gamma_k(V)$ (the maximal distance between vertices) is equal to k and we have $d(X, Y) = k$ for some $X, Y \in \mathcal{G}_k(V)$ only in the case when $X \cap Y = 0$. If $\dim V < 2k$, then the diameter of $\Gamma_k(V)$ is equal to $\dim V - k$.

Suppose that V is infinite-dimensional. For any $X, Y \in \mathcal{G}^k(V)$ the codimension of $X + Y$ is equal to the dimension of $X^0 \cap Y^0$ and the annihilator map is an isomorphism of $\Gamma^k(V)$ to $\Gamma_k(V^*)$. Therefore, the Grassmann graph $\Gamma^k(V)$ is connected for every natural k and the distance between $X, Y \in \mathcal{G}^k(V)$ in this graph is equal to k minus the codimension of $X + Y$.

In the proof of Proposition 2.11, we construct a geodesic of $\Gamma_k(V)$ from X to

Y, where every element contains $X \cap Y$. Now, we show that this property holds for every geodesic of $\Gamma_k(V)$ connecting X and Y.

Lemma 2.12 *If $X, X_1, \ldots, X_{i-1}, Y$ is a geodesic in $\Gamma_k(V)$, then*

$$X \cap Y \subset X \cap X_{i-1} \subset \cdots \subset X \cap X_1$$

and

$$X \cap Y \subset Y \cap X_1 \subset \cdots \subset Y \cap X_{i-1}.$$

Proof Since $X, X_1, \ldots, X_{i-1}, Y$ is a geodesic, we have $d(X, Y) = i$ and

$$d(X, X_j) = j, \quad d(X_j, Y) = i - j$$

for every $j \in \{1, \ldots, i-1\}$; in other words,

$$\dim(X \cap X_j) = k - j \quad \text{and} \quad \dim(Y \cap X_j) = k - i + j.$$

If $X \cap X_j$ does not contain $X \cap Y$, then the dimension of the intersection of these subspaces is less than $\dim(X \cap Y) = k - i$. This implies that $(X \cap X_j) \setminus Y$ contains a collection of $i - j + 1$ linearly independent vectors. Then

$$\dim X_j \geq i - j + 1 + \dim(Y \cap X_j) = (i - j + 1) + (k - i + j) = k + 1,$$

which is impossible. So, every $X \cap X_j$ contains $X \cap Y$.

Applying the same arguments to the geodesic $X, X_1, \ldots, X_j$ with $j \leq i - 1$, we establish that $X \cap X_j$ is contained in $X \cap X_l$ for every $l < j$.

To prove the second chain of inclusions, we consider the reversed geodesic $Y, X_{i-1}, \ldots, X_1, X$. □

Recall that an apartment of a Grassmannian of V consists of all elements of this Grassmannian spanned by all subsets of a certain basis of V.

Proposition 2.13 *Every geodesic of the graph $\Gamma_k(V)$ is contained in a certain apartment of $\mathcal{G}_k(V)$.*

Proof Let $X, X_1, \ldots, X_{i-1}, Y$ be a geodesic in $\Gamma_k(H)$. Then

$$\dim(X \cap X_j) = k - j \quad \text{and} \quad \dim(Y \cap X_j) = k - i + j$$

for every $j \in \{1, \ldots, i-1\}$. Lemma 2.12 implies the existence of vectors

$$x_1 \in X \setminus (X \cap X_1), \ x_2 \in (X \cap X_1) \setminus (X \cap X_2), \ \ldots, \ x_i \in (X \cap X_{i-1}) \setminus (X \cap Y)$$

and

$$y_1 \in Y \setminus (Y \cap X_{i-1}), \ y_2 \in (Y \cap X_{i-1}) \setminus (Y \cap X_{i-2}), \ \ldots, \ y_i \in (Y \cap X_1) \setminus (X \cap Y).$$

The vectors are linearly independent and every element from the geodesic is spanned by $X \cap Y$ and some of these vectors. Consider a basis of V containing $x_1, y_1, \ldots, x_i, y_i$ and such that $X \cap Y$ is spanned by subset of this basis. The associated apartment of $\mathcal{G}_k(V)$ is as required. □

If V is infinite-dimensional, then for any infinite cardinality $\alpha \le \dim V$ the Grassmann graphs $\Gamma_\alpha(V)$ and $\Gamma^\alpha(V)$ are not connected. The connected components in these graphs can be characterized as maximal subsets, where

$$\dim X/(X \cap Y) = \dim Y/(X \cap Y) < \infty$$

for any two elements X, Y. Using the above arguments, we can show that this number is the distance between X and Y.

A *clique* is a subset in the vertex set of a graph, where any two distinct vertices are adjacent. We describe maximal cliques in Grassmann graphs.

If X is a non-zero subspace of V whose codimension is greater than one, then the associated *star* $\mathcal{S}(X)$ consists of all subspaces of V containing X as a hyperplane. If Y is a proper subspace of V whose dimension is greater than one, then the associated *top* $\mathcal{T}(Y) = \mathcal{G}^1(Y)$ is formed by all hyperplanes of Y. All star and all top are cliques in some Grassmann graphs. Stars and tops of $\Gamma_k(V)$ are defined by $(k - 1)$-dimensional and $(k + 1)$-dimensional subspaces (respectively). Similarly, stars and tops of $\Gamma^k(V)$ are related to subspaces of codimension $k + 1$ and $k - 1$ (respectively). The annihilator map is an isomorphism between $\Gamma^k(V)$ and $\Gamma_k(V^*)$ which transfers stars to tops and tops to stars.

Proposition 2.14 *Every maximal clique in the Grassmann graph $\Gamma_\alpha(V)$ or $\Gamma^\alpha(V)$, $\alpha > 1$ is a star or a top.*

Proof Let $\mathcal{G}$ be a Grassmannian of V different from $\mathcal{G}_1(V)$ and $\mathcal{G}^1(V)$. We need to show that every clique of the associated Grassmann graph $\Gamma(\mathcal{G})$ is contained in a star or a top. We will use the following observation: if $X, Y \in \mathcal{G}$ are adjacent, then for every $Z \in \mathcal{G}$ adjacent to both X, Y we have

$$X \cap Y \subset Z \quad \text{or} \quad Z \subset X + Y$$

(it is possible that both inclusions hold). Suppose that $\mathcal{X}$ is a clique of $\Gamma(\mathcal{G})$ which is not contained in a star. Then there exist $X, Y, Z \in \mathcal{X}$ such that Z does not contain $X \cap Y$. In this case, Z is contained in $X + Y$. If $T \in \mathcal{G}$ is adjacent to X, Y and is not contained in $X + Y$, then it intersects $X + Y$ precisely in $X \cap Y$, which implies that T is not adjacent to Z. This means that $\mathcal{X}$ is a subset in the top $\mathcal{T}(X + Y)$. □

Let $\mathcal{G}$ be a Grassmannian of V whose elements are of infinite dimension and codimension and let C be a connected component in the associated Grassmann graph $\Gamma(\mathcal{G})$. For every positive integer i we denote by C_{-i} the set of all $Y \in \mathcal{G}$ such that Y is a subspace of codimension i in a certain element of C and we write C_{+i} for the set of all elements of $\mathcal{G}$ which contain some elements of C as subspaces of codimension i. We set $C_0 = C$ and denote by $C_{\pm}$ the union of all C_i, $i \in \mathbb{Z}$. If $X \in C_i$ for a certain $i \in \mathbb{Z}$, then every hyperplane of X belongs to C_{i-1} and every subspace containing X as a hyperplane is an element of C_{i+1}. An easy verification shows that for each $i \in \mathbb{Z}$ the following assertions are fulfilled:

- every C_i is a connected component of $\Gamma(\mathcal{G})$,
- a star $\mathcal{S}(X)$ is contained in C_i if and only if $X \in C_{i-1}$,
- a top $\mathcal{T}(Y)$ is contained in C_i if and only if $Y \in C_{i+1}$.

For any $X, Y \in C_{\pm}$ we have $X \cap Y \in C_{\pm}$ and $X + Y \in C_{\pm}$. Therefore, the partially ordered set $(C_{\pm}, \subset)$ is a lattice. This lattice is unbounded.

In what follows we will need some information concerning the intersections of maximal cliques. Let C be a connected component in $\Gamma_\alpha(V)$ or $\Gamma^\alpha(V)$ (if α is finite, then C coincides with the Grassmann graph).

The intersection of two distinct stars $\mathcal{S}(X), \mathcal{S}(Y) \subset C$ is empty or it consists of the single element $X + Y$. The second possibility is realized if and only if X, Y are adjacent (if the Grassmann graph is $\Gamma_k(V)$ or $\Gamma^k(V)$, then X, Y are adjacent elements of $\mathcal{G}_{k-1}(V)$ or $\mathcal{G}^{k+1}(V)$, respectively; in the case when the Grassmann graph is formed by subspaces of infinite dimension and codimension, X and Y are adjacent elements of C_{-1}).

Similarly, the intersection of two distinct tops $\mathcal{T}(X), \mathcal{T}(Y) \subset C$ is empty or it consists of the single element $X \cap Y$. As above, the second possibility is realized if and only if X, Y are adjacent (if the Grassmann graph is $\Gamma_k(V)$ or $\Gamma^k(V)$, then X, Y are adjacent elements of $\mathcal{G}_{k+1}(V)$ or $\mathcal{G}^{k-1}(V)$, respectively; and X, Y are adjacent elements of C_{+1} if the Grassmann graph is formed by subspaces of infinite dimension and codimension).

The intersection of a star $\mathcal{S}(X) \subset C$ and a top $\mathcal{T}(Y) \subset C$ is empty or it consists of all $Z \in C$ satisfying $X \subset Z \subset Y$. In the second case, the intersection is called a *line*. The star $\mathcal{S}(X)$ (together with all lines contained in it) is a projective space isomorphic to $\Pi_{V/X}$. The top $\mathcal{T}(Y)$ (together with all lines contained in it) is the dual projective space Π_Y^* which is isomorphic to Π_{Y^*}.

2.4 Automorphisms of Grassmann Graphs

Let V be as in the previous section. The restriction of every automorphism of the lattice $\mathcal{L}(V)$ to any Grassmannian of V is an automorphism of the corresponding Grassmann graph. If dim $V = 2k$, then the restriction of every anti-automorphism of $\mathcal{L}(V)$ to $\mathcal{G}_k(V)$ is an automorphism of $\Gamma_k(V)$.

First of all, we consider the classic Chow's theorem which originally was proved for Grassmannians of finite-dimensional vector spaces.

Theorem 2.15 (Chow [13])   *Let f be an automorphism of the Grassmann graph $\Gamma_k(V)$, where k is a natural number greater than one. In the case when* dim $V = n$ *is finite, we also require that $k < n - 1$. Then f can be uniquely extended to an automorphism of the lattice $\mathcal{L}(V)$ or* dim $V = 2k$ *and f is uniquely extendable to an anti-automorphism of the lattice $\mathcal{L}(V)$.*

Proof   It is clear that f and f^{-1} send maximal cliques (stars and tops) to maximal cliques. Since the intersection of two distinct maximal cliques is empty or a one-element set or a line, f and f^{-1} transfer lines to lines. Recall that every maximal clique of $\Gamma_k(V)$ (together with all lines contained in it) is a projective space. Therefore, the restriction of f to a maximal clique X is an isomorphism of the corresponding projective space to the projective space associated to $f(X)$.

Suppose that f sends a star $\mathcal{S}(X)$ to a top $\mathcal{T}(Y)$. Then it induces an isomorphism of $\Pi_{V/X}$ to Π_{Y^*} and $\dim(V/X) = \dim Y^*$. On the other hand, we have

$$\dim(V/X) = \dim V - k + 1 \quad \text{and} \quad \dim Y^* = \dim Y = k + 1,$$

which implies that dim $V = 2k$. For every $Y' \in \mathcal{G}_{k+1}(V)$ containing X the intersection of $\mathcal{S}(X)$ and $\mathcal{T}(Y')$ is a line, and the same holds for the intersection of $f(\mathcal{S}(X)) = \mathcal{T}(Y)$ and $f(\mathcal{T}(Y'))$. Thus $f(\mathcal{T}(Y'))$ is a star. If $X' \in \mathcal{G}_{k-1}(V)$ is adjacent to X, then we take $Y' \in \mathcal{G}_{k+1}(V)$ containing both X, X'. Using the above arguments, we establish that $f(\mathcal{S}(X'))$ is a top. The connectedness of the graph $\Gamma_{k-1}(V)$ guarantees that f transfers every star to a top. Similarly, we show that f sends tops to stars.

Therefore, one of the following possibilities is realized:

(A) f transfers stars and tops to stars and tops (respectively),
(B) dim $V = 2k$ and f sends stars to tops and tops to stars.

In the case (A), f induces a bijective transformation f_{k-1} of $\mathcal{G}_{k-1}(V)$ such that

$$f(\mathcal{S}(X)) = \mathcal{S}(f_{k-1}(X))$$

for every $X \in \mathcal{G}_{k-1}(V)$; in other words, for $X \in \mathcal{G}_{k-1}(V)$ and $Y \in \mathcal{G}_k(V)$ we have

$$X \subset Y \iff f_{k-1}(X) \subset f(Y).$$

The transformation f_{k-1} is an automorphism of $\Gamma_{k-1}(V)$ (two distinct stars in

$\Gamma_k(V)$ have a non-empty intersection if and only if the corresponding elements of $\mathcal{G}_{k-1}(V)$ are adjacent). Also, f_{k-1} transfers tops to tops (since it is induced by f), i.e. f_{k-1} sends stars to stars if $k \geq 3$. Step by step, we construct a sequence

$$f = f_k, f_{k-1}, \ldots, f_1,$$

where every f_i is an automorphism of $\Gamma_i(V)$. If $i, j \in \{1, \ldots, k\}$ and $i < j$, then for $X \in \mathcal{G}_i(V)$ and $Y \in \mathcal{G}_j(V)$ we have

$$X \subset Y \iff f_i(X) \subset f_j(Y).$$

In particular, f_1 is an automorphism of the projective space Π_V. Hence f_1 is induced by a semilinear automorphism L of V (unique up to a scalar multiple). Since $P \in \mathcal{G}_1(V)$ is contained in $X \in \mathcal{G}_k(V)$ if and only if $f_1(P) = L(P)$ is contained in $f(X)$, we have $f(X) = L(X)$ for every $X \in \mathcal{G}_k(V)$ and f can be uniquely extended to an automorphism of the lattice $\mathcal{L}(V)$.

In the case (B), the composition of f and the annihilator map is an isomorphism of $\Gamma_k(V)$ to $\Gamma_k(V^*)$ transferring stars to stars and tops to tops. By the arguments given above, it can be uniquely extended to an isomorphism between the lattices $\mathcal{L}(V)$ and $\mathcal{L}(V^*)$, which implies that f is uniquely extendable to an anti-automorphism of the lattice $\mathcal{L}(V)$. $\square$

Corollary 2.16 *Suppose that V is infinite-dimensional. Let f be an automorphism of the Grassmann graph $\Gamma^k(V)$, where k is a natural number greater than one. Then there exists a semilinear automorphism L of V^* unique up to a scalar multiple such that*

$$f(X) = {}^0L(X^0)$$

for every $X \in \mathcal{G}^k(V)$.

Proof Consider the automorphism of $\Gamma_k(V^*)$ transferring every $Y \in \mathcal{G}_k(V^*)$ to $f(^0Y)^0$. By Theorem 2.15, it is induced by a semilinear automorphism L of V^* (such a semilinear automorphism is unique up to a scalar multiple). An easy verification shows that L is as required. $\square$

Remark 2.17 An *isometric embedding* of a connected graph Γ to a connected graph Γ' is an injection of the vertex set of Γ to the vertex set of Γ' which preserves the distance between any pair of vertices. There is a description of isometric embeddings of Grassmann graphs formed by finite-dimensional subspaces. It is based on semilinear maps of special type associated to not necessarily surjective homomorphisms of division rings. The statement is proved for Grassmann graphs of finite-dimensional vector spaces only [45, Chapter 3]; but all arguments work for Grassmann graphs formed by finite-dimensional subspaces of vector spaces of arbitrary dimension. A modification of these arguments will be exploited in Chapter 4.

Now, we consider a Grassmannian $\mathcal{G}$ formed by subspaces of infinite dimension and codimension. It was observed in [8] that there are automorphisms of the Grassmann graph $\Gamma(\mathcal{G})$ non-extendable to automorphisms of the lattice $\mathcal{L}(V)$. This is related to the fact that $\Gamma(\mathcal{G})$ is non-connected.

Example 2.18   Let C be a connected component of $\Gamma(\mathcal{G})$ and let X, X' be distinct elements of C. Every semilinear automorphism L of V transferring X to X' induces a bijective transformation of C. We assert that the automorphism f of $\Gamma(\mathcal{G})$ which sends every $Y \in C$ to $L(Y)$ and leaves fixed every element of $\mathcal{G} \setminus C$ cannot be extended to an automorphism of $\mathcal{L}(V)$. First, we show that every 1-dimensional subspace $P \subset V$ can be presented as the intersection of some elements from $\mathcal{G} \setminus C$. We take any $Y \in \mathcal{G} \setminus C$ which does not contain P and such that $Y/(Z \cap Y)$ is infinite-dimensional for a certain $Z \in C$. Then for every hyperplane H of Y the subspace $H + P$ belongs to $\mathcal{G} \setminus C$ and the intersection of all such subspaces coincides with P. Suppose that f is induced by a semilinear automorphism S of V. Then S sends every element of $\mathcal{G} \setminus C$ to itself and we have $S(P) = P$ for every 1-dimensional subspace $P \subset V$. By Lemma 2.2, S is a scalar multiple of the identity, i.e. f is the identity and we get a contradiction.

Theorem 2.19 (Plevnik [52])   *Let $\mathcal{G}$ be a Grassmannian of V formed by subspaces of infinite dimension and codimension. Let also C and C' be connected components of the Grassmann graph $\Gamma(\mathcal{G})$. Suppose that f is the restriction of an automorphism of $\Gamma(\mathcal{G})$ to C and $f(C) = C'$. Then one of the following possibilities is realized:*

- *f can be uniquely extended to an isomorphism between the lattices $C_{\pm}$ and $C'_{\pm}$.*
- *f can be uniquely extended to an anti-isomorphism of the lattice $C_{\pm}$ to the lattice $C'_{\pm}$. Then $\mathcal{G} = \mathcal{G}_{\alpha}(V)$ and $\dim V = \beta^{\alpha}$, where β is the cardinality of the associated division ring.*

Proof   As in the proof of Theorem 2.15, one of the following possibilities is realized:

(A) f sends stars to stars and tops to tops,
(B) f sends stars to tops and tops to stars.

In the case (A), f induces bijections

$$f_{-1} : C_{-1} \to C'_{-1} \quad \text{and} \quad f_{+1} : C_{+1} \to C'_{+1}$$

which are isomorphisms between the restrictions of the graph $\Gamma(\mathcal{G})$ to the corresponding connected components. Observe that f_{-1} and f_{+1} transfer stars to stars and tops to tops. Step by step, we construct a sequence $\{f_i\}_{i\in\mathbb{Z}}$, where every f_i is a bijection of C_i to C_i' ($f_0 = f$). If $i, j \in \mathbb{Z}$ and $i < j$, then for $X \in C_i$ and $Y \in C_j$ we have

$$X \subset Y \iff f_i(X) \subset f_j(Y).$$

The union of all f_i is an isomorphism of the lattice $C_{\pm}$ to the lattice $C_{\pm}'$.

The extension of f to a lattice isomorphism is unique. Indeed, if F_1 and F_2 are such extensions and $X \in C_{\pm}$, then $F_1(X)$ is incident to $X' \in C'$ if and only if $F_2(X)$ is incident to X', which implies that $F_1(X) = F_2(X)$.

In the case (B), f induces bijections

$$f_{-1} : C_{-1} \to C_{+1}' \quad \text{and} \quad f_{+1} : C_{+1} \to C_{-1}'.$$

These are isomorphisms between the restrictions of the graph $\Gamma(\mathcal{G})$ to the corresponding connected components which send stars to tops and tops to stars. We construct a sequence $\{f_i\}_{i\in\mathbb{Z}}$, where every f_i is a bijection of C_i to C_{-i}'. If $i, j \in \mathbb{Z}$ and $i < j$, then for $X \in C_i$ and $Y \in C_j$ we have

$$X \subset Y \iff f_j(Y) \subset f_i(X).$$

The union of all f_i is an anti-isomorphism of $C_{\pm}$ to $C_{\pm}'$. As above, we establish that this extension is unique.

Suppose that f sends the star $\mathcal{S}(X)$, $X \in C_{-1}$ to the top $\mathcal{T}(Y)$, $Y \in C_1'$ and F is the extension of f to an anti-automorphism of $C_{\pm}$ to $C_{\pm}'$. The restriction of F to the set of all elements of $C_{\pm}$ containing X can be considered as an anti-isomorphism of the lattice $\mathcal{L}_{\mathrm{fin}}(V/X)$ to the lattice $\mathcal{L}^{\mathrm{fin}}(Y)$. Using the annihilator map, we obtain an isomorphism between the lattices $\mathcal{L}_{\mathrm{fin}}(V/X)$ and $\mathcal{L}_{\mathrm{fin}}(Y^*)$. It induces an isomorphism between the projective spaces $\Pi_{V/X}$ and Π_{Y^*}. Therefore,

$$\dim(V/X) = \dim Y^*.$$

Let α be the dimension of elements from $\mathcal{G}$. By Theorem 1.9, $\dim Y^* = \beta^\alpha$, where β is the cardinality of the associated division ring. Then $\dim(V/X) = \beta^\alpha$, which implies that $\dim V = \beta^\alpha$ and $\mathcal{G} = \mathcal{G}_\alpha(V)$. □

A description of all isomorphisms and anti-isomorphisms between the lattices $C_{\pm}$ and $C_{\pm}'$ is an open problem.

Remark 2.20 This problem is considered in [52]. One of the results says that every isomorphism of $C_{\pm}$ to $C_{\pm}'$ can be obtained from a pair of semilinear isomorphisms of special type; but this fails. We briefly describe this construction.

Let $U_0 \in C_\pm$ and $V_0 \in C'_\pm$. Suppose that U_1 and V_1 are complements of U_0 and V_0, respectively, i.e.

$$U_0 \oplus U_1 = V = V_0 \oplus V_1.$$

For any finite-dimensional subspaces $Z \subset U_1^0$ and $U' \subset U_1$ the subspace

$$(U_0 \cap {}^0 Z) \oplus U'$$

belongs to $C_\pm$. The set of all such elements of $C_\pm$ will be denoted by $\mathcal{Z}$. Similarly, any pair formed by a finite-dimensional subspace of V_1^0 and a finite-dimensional subspace of V_1 defines an element of $C'_\pm$ and we write $\mathcal{Z}'$ for the set of all such elements. Any two semilinear isomorphisms $A : U_1^0 \to V_1^0$ and $B : U_1 \to V_1$ define the bijection $\rho_{A,B} : \mathcal{Z} \to \mathcal{Z}'$ as follows:

$$\rho_{A,B}((U_0 \cap {}^0 Z) \oplus U') = (V_0 \cap {}^0 A(Z)) \oplus B(U').$$

The result states that every isomorphism of $C_\pm$ to $C'_\pm$ is a map of such type. However, $\mathcal{Z}$ and $\mathcal{Z}'$ are proper subsets of C and C' (respectively). Indeed, for every hyperplane H in U_0 the subspace $H + U_1$ is a hyperplane of V and we take any 1-dimensional subspace $P \not\subset U_0$ satisfying $H + U_1 + P = V$. Then $H + P$ belongs to $C_\pm \setminus \mathcal{Z}$.

Now, we explain why the above-mentioned problem is hard. Let f be an isomorphism of the lattice $C_\pm$ to the lattice $C'_\pm$. For every $X \in C_\pm$ we denote by f_X the restriction of f to $\mathcal{L}^{\text{fin}}(X)$. This is an isomorphism to the lattice $\mathcal{L}^{\text{fin}}(f(X))$. It induces an isomorphism between the dual projective spaces Π_X^* and $\Pi_{f(X)}^*$. Using this fact, we establish the existence of a semilinear isomorphism $S_X : X^* \to f(X)^*$ such that

$$f_X(Y) = {}^0 S_X(Y^0)$$

for every $Y \in \mathcal{L}^{\text{fin}}(X)$. If every S_X is the contragradient of a certain semilinear isomorphism $L_X : X \to f(X)$, then

$$f_X(Y) = L_X(Y)$$

for all $Y \in \mathcal{L}^{\text{fin}}(X)$ (see the end of Section 2.1). In this case, we can construct a semilinear automorphism of V whose restriction to every $X \in C_\pm$ is a scalar multiple of L_X and show that f is induced by this semilinear automorphism (see Section 2.6 for the details). However, we work with infinite-dimensional vector spaces and S_X is not necessarily adjoint to a semilinear isomorphism.

2.5 Complementary Preserving Transformations

Let $\mathcal{G}$ be the Grassmannian formed by subspaces of V whose dimension is equal to the codimension. Then one of the following possibilities is realized:

- $\dim V = 2k$ is finite and $\mathcal{G} = \mathcal{G}_k(V)$,
- $\dim V = \alpha$ is infinite and $\mathcal{G} = \mathcal{G}_\alpha(V) = \mathcal{G}^\alpha(V)$.

Observe that for every $X \in \mathcal{G}$ all complements of X also belong to $\mathcal{G}$. Denote by $\Delta(V)$ the graph whose vertex set is $\mathcal{G}$ and two elements of $\mathcal{G}$ are connected by an edge in this graph if they are complementary subspaces.

The restriction of every automorphism of the lattice $\mathcal{L}(V)$ to $\mathcal{G}$ is an automorphism of the graph $\Delta(V)$. If $\dim V = 2k$, then the restriction of every anti-automorphism of $\mathcal{L}(V)$ to $\mathcal{G} = \mathcal{G}_k(V)$ is an automorphism of $\Delta(V)$. In this section, we determine all automorphisms of this graph. We will use the following characterization of the adjacency relation in terms of complementaries.

Theorem 2.21 (Blunck and Havlicek [8]) *Let X_1 and X_2 be distinct elements in a certain Grassmannian $\mathcal{G}$ of V. The following conditions are equivalent:*

(1) *X_1 and X_2 are adjacent elements of $\mathcal{G}$,*
(2) *the Grassmannian $\mathcal{G}$ contains an element X distinct from X_1, X_2 and such that every complement of X is also a complement of at least one of X_i.*

By Theorem 2.21, every automorphism of the graph $\Delta(V)$ is an automorphism of the Grassmann graph $\Gamma(\mathcal{G})$. If $\dim V = 2k$, then Theorem 2.15 implies that every automorphism of $\Delta(V)$ can be uniquely extended to an automorphism or anti-automorphism of the lattice $\mathcal{L}(V)$.

Proof of Theorem 2.21 (1) $\Longrightarrow$ (2) If X_1 and X_2 are adjacent, then we consider the unique line of $\mathcal{G}$ containing X_1 and X_2, i.e. the set formed by all $X \in \mathcal{G}$ satisfying

$$X_1 \cap X_2 \subset X \subset X_1 + X_2.$$

Every such $X \neq X_1, X_2$ is as required. Indeed, if Y is a complement of X, then Y intersects $X_1 + X_2$ in a 1-dimensional subspace P which is not contained in $X_1 \cap X_2$. This means that $Y \cap X_i = 0$ for at least one of $i \in \{1, 2\}$. For such i we have $X + P = X_i + P$ and Y is a complement of X_i.

(2) $\Longrightarrow$ (1) If $\mathcal{G}$ coincides with $\mathcal{G}_1(V)$ or $\mathcal{G}^1(V)$, then the implication is trivial (in this case, any two distinct elements in $\mathcal{G}$ are adjacent). Suppose that $\mathcal{G}$ is $\mathcal{G}_\beta(V)$ or $\mathcal{G}^\beta(V)$ with $\beta > 1$.

Let X be an element of $\mathcal{G}$ satisfying (2). First of all, we observe that for any 1-dimensional subspaces $P_1 \subset X_1$ and $P_2 \subset X_2$ the sum $P_1 + P_2$ has a non-zero intersection with X (if this fails, then there is a complement of X containing $P_1 + P_2$; by our assumption, this subspace is a complement of X_1 or X_2, which is impossible). If $X_1 \cap X_2$ is non-empty, then we apply the above observation to $P_1 = P_2 \subset X_1 \cap X_2$ and get the inclusion

$$X_1 \cap X_2 \subset X. \tag{2.2}$$

Our next step is to prove that

$$X \subset X_1 + X_2. \tag{2.3}$$

This is trivial if $X_1 + X_2$ coincides with V. In the case when $X_1 + X_2 \neq V$, it is sufficient to show that X is contained in every hyperplane $H \in \mathcal{G}^1(V)$ containing $X_1 + X_2$ (since the intersection of all such hyperplanes coincides with $X_1 + X_2$). If a hyperplane $H \in \mathcal{G}^1(V)$ does not contain X, then it contains a certain complement Y of X. The inclusion $X_1 + X_2 \subset H$ shows that Y is not a complement of X_i for each $i \in \{1, 2\}$. The latter contradicts our assumption.

The subspaces X_1 and X_2 are not incident. Indeed, if X_1 is contained in X_2, then (2.2) and (2.3) imply that $X_1 \subset X \subset X_2$. Since X, X_1, X_2 are mutually distinct, every complement of X is not a complement of X_i for each $i \in \{1, 2\}$. Similarly, we establish that X_2 is not contained in X_1.

Now, we show that neither X_1 nor X_2 is contained in X. Suppose, for example, that X_1 is contained in X. Since X_1 and X_2 are not incident, there is a hyperplane $H \in \mathcal{G}^1(V)$ which contains X_2 and does not contain X_1. Then

$$V = X_1 + H \subset X + H,$$

i.e. $X + H$ coincides with V. This means that H contains a certain complement Y of X. Since X_1 is a proper subspace of X, the subspace Y is not a complement of X_1. Also, Y is not a complement of X_2 (because Y and X_2 both are contained in H). We get a contradiction, which implies that X does not contain X_1. By the same arguments, X does not contain X_2.

We take a 1-dimensional subspace $P_1 \subset X_1$ which is not contained in X (this is possible since $X_1 \not\subset X$). If x_2 is a non-zero vector of X_2, then $\langle x_2 \rangle \neq P_1$ (otherwise, $P_1 \subset X_1 \cap X_2 \subset X$ by (2.2), which is impossible) and $P_1 + \langle x_2 \rangle$ has a non-zero intersection with X. Therefore, $X_2 \subset X + P_1$. Similarly, we have $X_1 \subset X + P_2$ for a 1-dimensional subspace $P_2 \subset X_2$ which is not contained in X. Then $X_1 + X_2$ is contained in $X + P_1 + P_2$ and (2.3) shows that

$$X_1 + X_2 = X + P_1 + P_2.$$

Since $P_1 + P_2$ has a non-zero intersection with X, we obtain that

$$X + P_1 = X_1 + X_2 = X + P_2$$

and X is a hyperplane of $X_1 + X_2$.

If Y is a complement of $X_1 + X_2$, then the latter equality shows that $Y + P_1$ is a complement of X. The subspace $Y + P_1$ cannot be a complement of X_1 (since P_1 is contained in X_1), hence it is a complement of X_2. Then

$$Y + (X_1 + X_2) = V = Y + (X_2 + P_1)$$

and the inclusion $X_2 + P_1 \subset X_1 + X_2$ shows that

$$X_1 + X_2 = X_2 + P_1.$$

Similarly, we establish that

$$X_1 + X_2 = X_1 + P_2.$$

Therefore, each X_i is a hyperplane of $X_1 + X_2$, which means that X_1 and X_2 are adjacent. $\qquad\square$

From this moment, we suppose that dim $V = \alpha$ is infinite and $G = G_\alpha(V)$.

Theorem 2.22 (Blunck and Havlicek [7]) *If V is infinite-dimensional, then the diameter of the graph $\Delta(V)$ is equal to three.*

Proof Suppose that $X, Y \in G_\alpha(V)$ are non-complementary subspaces. We put $X_1 = X \cap Y$ and take subspaces X_2, X_3 such that

$$X_1 \oplus X_2 = Y \quad \text{and} \quad X_1 \oplus X_3 = X.$$

There is a complement Z of X containing X_2. Let X_4 be a subspace satisfying $X_2 \oplus X_4 = Z$. Then

$$V = X_1 \oplus X_2 \oplus X_3 \oplus X_4$$

and there are two possibilities:

(1) The subspaces X_2 and X_3 are of the same dimension. Then there is a linear isomorphism $L : X_2 \to X_3$. We define

$$X' = \{x + L(x) : x \in X_2\}$$

and obtain that

$$V = X_1 \oplus X_2 \oplus X' \oplus X_4 = X_1 \oplus X' \oplus X_3 \oplus X_4.$$

This means that $X' + X_4$ is a complement of both $X = X_1 + X_3$ and $Y = X_1 + X_2$, i.e. the distance between X and Y in the graph $\Delta(V)$ is equal to two.

(2) The dimensions of X_2 and X_3 are distinct. We assert that dim $X_1 = \alpha$. Indeed, if this fails, then the equality

$$\max\{\dim X_1, \dim X_3\} = \dim X = \alpha = \dim Y = \max\{\dim X_1, \dim X_2\}$$

implies that $\dim X_2 = \alpha = \dim X_3$, which is impossible. Since $Z = X_2 + X_4$ and $X_3 + X_4$ belong to $\mathcal{G}_\alpha(V)$ (the second subspace is a complement of Y), the same arguments show that $\dim X_4 = \alpha$. Then X_1 and X_4 are isomorphic. As in the previous case, we construct a subspace X' such that

$$V = X_1 \oplus X_2 \oplus X_3 \oplus X' = X' \oplus X_2 \oplus X_3 \oplus X_4.$$

Therefore, $X' + X_3$ is a complement of both $Z = X_2 + X_4$ and $Y = X_1 + X_2$. Since Z is a complement of X, the distance between X and Y in the graph $\Delta(V)$ is not greater than three.

So, the distance between any two vertices of $\Delta(V)$ is not greater than three. On the other hand, the graph $\Delta(V)$ contains pairs of vertices with distance greater than two. Consider, for example, any two incident elements of $\mathcal{G}_\alpha(V)$.

□

2.6 Apartments Preserving Transformations

As above, we suppose that $\mathcal{G}$ is a Grassmannian of a vector space V. If B is a basis of V, then the *apartment* of $\mathcal{G}$ associated to the basis B consists of all elements of $\mathcal{G}$ spanned by subsets of B. It was noted above that for any two elements of $\mathcal{G}$ there is an apartment containing them (Proposition 1.7) and the apartments of $\mathcal{G}$ defined by bases B and B' are coincident if and only if the vectors from B' are scalar multiples of the vectors from B. We say that an apartment $\mathcal{A} \subset \mathcal{G}$ is associated to an apartment of another Grassmannian of V if these apartments are defined by the same basis.

If $\dim V = n$ is finite, then the annihilator map transfers the apartment of $\mathcal{G}_k(V)$ associated to a basis $e_1, \ldots, e_n$ to the apartment of $\mathcal{G}_{n-k}(V^*)$ defined by the dual basis $e_1^*, \ldots, e_n^*$. In the case when V is infinite-dimensional, the annihilator map sends apartments of $\mathcal{G}^k(V)$ to proper subsets of apartments of $\mathcal{G}_k(V^*)$; see Remark 1.11.

Remark 2.23 Suppose that V is a vector space over a field and consider the exterior k-product $\wedge^k V$ (which is defined for an arbitrary, not necessarily finite-dimensional, vector space). This is the vector space (over the same field) whose elements are linear combinations of so-called k-*vectors* $x_1 \wedge \cdots \wedge x_k$, where $x_1, \ldots, x_k$ are linearly independent vectors from V (see [60] for the precise definition). If $\{e_i\}_{i \in I}$ is a basis of the vector space V, then all k-vectors of type $e_{i_1} \wedge \cdots \wedge e_{i_k}$, where $i_1, \ldots, i_k$ are mutually distinct elements of I, form a basis of the vector space $\wedge^k V$. Such bases of $\wedge^k V$ are said to be *regular*. If $\dim V = n$

is finite, then

$$\dim(\wedge^k V) = \binom{n}{k}.$$

If vectors $x_1, \ldots, x_k$ and $y_1, \ldots, y_k$ span the same k-dimensional subspace of V, then

$$y_1 \wedge \cdots \wedge y_k = \det(M)\, x_1 \wedge \cdots \wedge x_k,$$

where M is the matrix of decomposition of $y_1, \ldots, y_k$ with respect to the basis $x_1, \ldots, x_k$. Therefore, the k-dimensional subspace of V spanned by vectors $x_1, \ldots, x_k$ can be naturally identified with the 1-dimensional subspace of $\wedge^k V$ containing the k-vector $x_1 \wedge \cdots \wedge x_k$. We get an injective map of $\mathcal{G}_k(V)$ to $\mathcal{G}_1(\wedge^k V)$ whose image consists of all 1-dimensional subspaces of $\wedge^k V$ containing k-vectors. This map is known as the *Plücker embedding*. It transfers every line of $\mathcal{G}_k(V)$ to a line of the projective space $\Pi_{\wedge^k V}$. The apartment of $\mathcal{G}_k(V)$ defined by a basis B goes to the apartment of $\mathcal{G}_1(\wedge^k V)$ defined by the regular basis of $\wedge^k V$ corresponding to B.

Let $\mathcal{G}$ be a Grassmannian of V. Recall that $\mathcal{G}$ is $\mathcal{G}_\alpha(V)$ or $\mathcal{G}^\alpha(V)$, where α is a cardinality not greater than the dimension of V. The restriction of every automorphism of the lattice $\mathcal{L}(V)$ to $\mathcal{G}$ sends apartments to apartments. If $\dim V = 2k$ and $\mathcal{G} = \mathcal{G}_k(V)$, then the same holds for the restrictions of anti-automorphisms of $\mathcal{L}(V)$ to $\mathcal{G}$.

Theorem 2.24 (Pankov [47]) *Let f be a bijective transformation of $\mathcal{G}$ such that f and f^{-1} send apartments to apartments. Then f can be uniquely extended to an automorphism of the lattice $\mathcal{L}(V)$ or $\dim V = 2k$, $\mathcal{G} = \mathcal{G}_k(V)$ and f is uniquely extendable to an anti-automorphism of $\mathcal{L}(V)$.*

The proof of this theorem is split into three parts: the first is immediately below, the second follows Lemma 2.27 and the third is after Lemma 2.28.

Proof of Theorem 2.24 for the case when $\alpha = 1$ Suppose that $\mathcal{G} = \mathcal{G}_1(V)$. Recall that three points of a projective space are called *collinear* if there is a line containing them. Observe that three points of the projective space Π_V are non-collinear if and only if there is an apartment of $\mathcal{G}_1(V)$ containing them. This implies that f and f^{-1} send any triple of non-collinear points to a triple of non-collinear points. Then these maps transfer triples of collinear points to triples of collinear points, which means that f is an automorphism of Π_V.

Let $\mathcal{G} = \mathcal{G}^1(V)$. We claim that for every $P \in \mathcal{G}_1(V)$ there is a unique $P' \in \mathcal{G}_1(V)$ such that $X \in \mathcal{G}^1(V)$ contains P if and only if $f(X)$ contains P'.

For every $P \in \mathcal{G}_1(V)$ we take an apartment $\mathcal{A}_1 \subset \mathcal{G}_1(V)$ containing P and denote by $\mathcal{A}$ the associated apartment of $\mathcal{G}^1(V)$. There is a unique element

$X \in \mathcal{A}$ which does not contain P. Since $f(\mathcal{A})$ is an apartment of $\mathcal{G}^1(V)$, the intersection of all elements from $f(\mathcal{A}) \setminus \{f(X)\}$ is a 1-dimensional subspace P'. If a hyperplane $Y \in \mathcal{G}^1(V)$ does not contain P, then

$$(\mathcal{A} \setminus \{X\}) \cup \{Y\} \tag{2.4}$$

is an apartment of $\mathcal{G}^1(V)$. Indeed, for every $Q \in \mathcal{A}_1 \setminus \{P\}$ the hyperplane Y intersects the 2-dimensional subspace $P+Q$ in a certain $Q' \in \mathcal{G}_1(V)$ distinct from P, all such Q' together with P form an apartment of $\mathcal{G}_1(V)$ and the associated apartment of $\mathcal{G}^1(V)$ is (2.4). Therefore,

$$f((\mathcal{A} \setminus \{X\}) \cup \{Y\}) = (f(\mathcal{A}) \setminus \{f(X)\}) \cup \{f(Y)\}$$

is an apartment of $\mathcal{G}^1(V)$, which implies that $f(Y)$ does not contain P'. Applying the same arguments to f^{-1}, we establish that a hyperplane $Y \in \mathcal{G}^1(V)$ does not contain P if and only if $f(Y)$ does not contain P', i.e. P' is as required. The uniqueness of P' follows from the fact that for any two distinct elements of $\mathcal{G}_1(V)$ there is a hyperplane of V which contains one of these elements and does not contain the other.

So, there is a transformation g of $\mathcal{G}_1(V)$ such that

$$P \subset X \iff g(P) \subset f(X)$$

for $P \in \mathcal{G}_1(V)$ and $X \in \mathcal{G}^1(V)$. It is clear that g is bijective. Observe that $P \in \mathcal{G}_1(V)$ is the intersection of all elements from a subset $\mathcal{X} \subset \mathcal{G}^1(V)$ if and only if $g(P)$ is the intersection of all elements from $f(\mathcal{X})$.

Assume that f sends an apartment $\mathcal{A}_1 \subset \mathcal{G}^1(V)$ to an apartment $\mathcal{A}_2 \subset \mathcal{G}^1(V)$. For $i = 1, 2$ we denote by $\mathcal{A}'_i$ the apartment of $\mathcal{G}_1(V)$ associated to $\mathcal{A}_i$. Since every $P \in \mathcal{A}'_i$ is the intersection of all elements of $\mathcal{A}_i$ containing P, we have $g(\mathcal{A}'_1) = \mathcal{A}'_2$. So, g transfers apartments to apartments. Similarly, we show that g^{-1} sends apartments to apartments. Then g is induced by a semilinear automorphism of V and this semilinear automorphism induces f.   □

From this moment, we suppose that $\alpha > 1$. Let $B = \{e_i\}_{i \in I}$ be a basis of V and let $\mathcal{A}$ be the associated apartment of $\mathcal{G}$. For every $i \in I$ we denote by $\mathcal{A}(+i)$ the subset of $\mathcal{A}$ consisting of all elements which contain e_i and write $\mathcal{A}(-i)$ for the subset of $\mathcal{A}$ formed by all elements which do not contain e_i. We say that $\mathcal{A}(+i)$ and $\mathcal{A}(-i)$ are *simple subsets of first* and *second type*, respectively. For distinct $i, j \in I$ we define

$$\mathcal{A}(+i, +j) = \mathcal{A}(+i) \cap \mathcal{A}(+j),$$

$$\mathcal{A}(+i, -j) = \mathcal{A}(+i) \cap \mathcal{A}(-j).$$

A subset of $\mathcal{A}$ is called *inexact* if there is an apartment of $\mathcal{G}$ distinct from $\mathcal{A}$ and containing this subset. For any distinct $i, j \in I$ the subset

$$\mathcal{A}(+i, +j) \cup \mathcal{A}(-i) \tag{2.5}$$

is inexact; it is also contained in the apartment defined by the basis

$$(B \setminus \{e_i\}) \cup \{e_i'\},$$

where $e_i' = ae_i + be_j$ and a, b are non-zero scalars.

Lemma 2.25 *Every inexact subset of $\mathcal{A}$ is contained in a maximal inexact subset of $\mathcal{A}$ and each maximal inexact subset of $\mathcal{A}$ is a subset of type* (2.5).

Proof We need to show that every inexact subset $X \subset \mathcal{A}$ is contained in a subset of type (2.5). For every $i \in I$ we denote by S_i the intersection of all elements of X containing e_i (we assume that $S_i = 0$ if each element of X does not contain e_i). There is $i \in I$ such that $\dim S_i \neq 1$ (otherwise X is not inexact). If $S_i = 0$, then X is a subset of $\mathcal{A}(-i)$. In the case when $\dim S_i \geq 2$, we take any $e_j \in S_i$ such that $j \neq i$. Then X is contained in the set (2.5). Indeed, $X \in \mathcal{X}$ belongs to $\mathcal{A}(-i)$ if it does not contain e_i; if e_i belongs to X, then $e_j \in S_i \subset X$ and X is an element of $\mathcal{A}(+i, +j)$. $\quad\square$

A subset $X \subset \mathcal{A}$ is said to be *complementary* if $\mathcal{A} \setminus X$ is a maximal inexact subset. The complementary subset of $\mathcal{A}$ corresponding to the maximal inexact subset (2.5) is $\mathcal{A}(+i, -j)$.

For two distinct complementary subsets $\mathcal{A}(+i, -j)$ and $\mathcal{A}(+i', -j')$ one of the following possibilities is realized:

- $i = i'$ or $j = j'$,
- $i = j'$ or $j = i'$, which implies that the intersection of the complementary subsets is empty,
- $\{i, j\} \cap \{i', j'\} = \emptyset$.

In the first case, we say that the complementary subsets are *adjacent*. Observe that distinct complementary subsets $X, \mathcal{Y} \subset \mathcal{A}$ are adjacent if and only if their intersection is maximal, i.e. for any distinct complementary subsets $X', \mathcal{Y}' \subset \mathcal{A}$ satisfying

$$X \cap \mathcal{Y} \subset X' \cap \mathcal{Y}'$$

the inverse inclusion holds. It is not difficult to see that every maximal collection of mutually adjacent complementary subsets of $\mathcal{A}$ is

$$\{\mathcal{A}(+i, -j)\}_{j \in I \setminus \{i\}} \quad \text{or} \quad \{\mathcal{A}(+j, -i)\}_{j \in I \setminus \{i\}}$$

for a certain fixed $i \in I$. Note that $\mathcal{A}(+i)$ and $\mathcal{A}(-i)$ are the unions of all elements from the first and the second collection, respectively. Therefore, every simple subset of $\mathcal{A}$ can be characterized as the union of all elements from a maximal collection of mutually adjacent complementary subsets.

Let $\mathcal{A}'$ be the apartment of G defined by a basis $\{e'_i\}_{i \in I}$. The simple subsets of $\mathcal{A}'$ associated to the vector e'_i will be denoted by $\mathcal{A}'(+i)$ and $\mathcal{A}'(-i)$. A bijection $g : \mathcal{A} \to \mathcal{A}'$ is called *special* if g and g^{-1} transfer inexact subsets to inexact subsets.

Lemma 2.26 *If $g : \mathcal{A} \to \mathcal{A}'$ is a special bijection, then there is a bijective transformation $\delta : I \to I$ such that one of the following possibilities is realized:*

(1) $g(\mathcal{A}(+i)) = \mathcal{A}'(+\delta(i))$ and $g(\mathcal{A}(-i)) = \mathcal{A}'(-\delta(i))$ for all $i \in I$,

(2) $g(\mathcal{A}(+i)) = \mathcal{A}'(-\delta(i))$ and $g(\mathcal{A}(-i)) = \mathcal{A}'(+\delta(i))$ for all $i \in I$.

Proof It is clear that g and g^{-1} send maximal inexact subsets to maximal inexact subsets. This means that $X \subset \mathcal{A}$ is a complementary subset if and only if $g(X)$ is a complementary subset of $\mathcal{A}'$. Since two complementary subsets are adjacent only in the case when their intersection is maximal, g and g^{-1} transfer adjacent complementary subsets to adjacent complementary subsets. Every simple subset can be characterized as the union of all elements from a maximal collection of mutually adjacent complementary subsets. This implies that g and g^{-1} send simple subsets to simple subsets. Observe that two distinct simple subsets are of different types if and only if their intersection is empty or a complementary subset. Therefore, g and g^{-1} transfer simple subsets of different types to simple subsets of different types, i.e. g and g^{-1} preserve the types of all simple subsets or change the type of every simple subset. Since $\mathcal{A}(-i) = \mathcal{A} \setminus \mathcal{A}(+i)$, there is a bijection $\delta : I \to I$ such that the case (1) or (2) is realized. □

We say that $g : \mathcal{A} \to \mathcal{A}'$ is a special bijection of *first* or of *second type* if it satisfies (1) or (2), respectively.

Suppose that $X \in \mathcal{A}$ is spanned by all vectors e_i with $i \in J \subset I$. Then (1) implies that $g(X)$ is spanned by all e'_i with $i \in \delta(J)$. In the case (2), $g(X)$ is spanned by all e'_i with $i \in I \setminus \delta(J)$. Therefore, the second possibility can be realized only in the case when the dimension of X is equal to the codimension; in other words, $G = G_\alpha(V)$ and $\dim V = \alpha$ is infinite or $\dim V = 2\alpha$ is finite.

Recall that f is a bijective transformation of G such that f and f^{-1} send apartments to apartments.

Lemma 2.27 *The transformation f is an automorphism of the Grassmann graph $\Gamma(G)$.*

Proof Lemma 2.26 shows that every special bijection $g : \mathcal{A} \to \mathcal{A}'$ is an isomorphism between the restrictions of the Grassmann graph $\Gamma(\mathcal{G})$ to the apartments $\mathcal{A}$ and $\mathcal{A}'$. For any two elements of $\mathcal{G}$ there is an apartment containing them and the restriction of f to every apartment $\mathcal{A} \subset \mathcal{G}$ is a special bijection of $\mathcal{A}$ to $f(\mathcal{A})$. $\qquad\square$

Proof of Theorem 2.24 for the case when $1 < \alpha < \infty$ Suppose that α is finite and greater than one. If $\mathcal{G} = \mathcal{G}_\alpha(V)$, then Theorem 2.24 follows from Lemma 2.27 and Theorem 2.15.

Consider the case when V is infinite-dimensional and $\mathcal{G} = \mathcal{G}^\alpha(V)$. By Lemma 2.27 and Corollary 2.16, there is a semilinear automorphism S of V^* such that

$$f(X) = {}^0S(X^0)$$

for every $X \in \mathcal{G}^\alpha(V)$. Denote by g the bijective transformation of $\mathcal{G}^1(V)$ which sends every $H \in \mathcal{G}^1(V)$ to ${}^0S(H^0)$. Then

$$X \subset H \iff f(X) \subset g(H)$$

for $X \in \mathcal{G}^\alpha(V)$ and $H \in \mathcal{G}^1(V)$. A direct verification shows that $H \in \mathcal{G}^1(V)$ is the sum of some $X, Y \in \mathcal{G}^\alpha(V)$ if and only if $g(H) = f(X) + f(Y)$.

Assume that f sends an apartment $\mathcal{A}_1 \subset \mathcal{G}^\alpha(V)$ to an apartment $\mathcal{A}_2 \subset \mathcal{G}^\alpha(V)$. For $i = 1, 2$ we denote by $\mathcal{A}'_i$ the apartments of $\mathcal{G}^1(V)$ associated to $\mathcal{A}_i$. Every element of $\mathcal{A}'_i$ can be presented as the sum of two elements from $\mathcal{A}_i$. This means that $g(\mathcal{A}'_1) = \mathcal{A}'_2$. Therefore, g transfers apartments to apartments. The same arguments show that g^{-1} sends apartments to apartments. Then g is induced by a semilinear automorphism of V which also induces f. $\qquad\square$

Lemma 2.28 *If α is infinite, then f is an automorphism of the partially ordered set* $(\mathcal{G}, \subset)$.

Proof First, we show that the restriction of f to every apartment is a special bijection of first type.

It was noted above that every special bijection between apartments of $\mathcal{G}$ is of first type if $\alpha < \dim V$. Suppose that $\dim V = \alpha$ and $\mathcal{G} = \mathcal{G}_\alpha(V)$. By Lemma 2.27, f is an automorphism of $\Gamma(\mathcal{G})$; moreover, f transfers stars to stars and tops to tops. Indeed, if f sends a star $\mathcal{S}(X)$ to a top $\mathcal{T}(Y)$, then it induces an isomorphism between the projective spaces $\Pi_{V/X}$ and Π_{Y^*} (see the proof of Theorem 2.15) and the codimension of X is equal to the dimension of Y^*, which is impossible if X and Y are elements of $\mathcal{G}_\alpha(V)$ and $\dim V = \alpha$ (see Theorem 1.9). Let $\mathcal{A}$ be an apartment of $\mathcal{G}$. We take any $X \in \mathcal{A}$ and consider the star $\mathcal{S}(X)$. The transformation f sends $\mathcal{A} \cap \mathcal{S}(X)$ to the intersection of the apartment $f(\mathcal{A})$ with a certain star. If the restriction of f to $\mathcal{A}$ is a special bijection of second type, then $f(\mathcal{A} \cap S(X))$ is the intersection of $f(\mathcal{A})$ with a top and we get a contradiction.

Let $X, Y \in \mathcal{G}$. Consider an apartment $\mathcal{A} \subset \mathcal{G}$ containing X and Y. It was established above that the restriction of f to $\mathcal{A}$ is a special bijection of first type. Lemma 2.26 shows that X is contained in Y if and only if $f(X)$ is contained in $f(Y)$. □

Proof of Theorem 2.24 for the case when α is infinite   Suppose that α is an infinite cardinality. Let $X \in \mathcal{G}$. It follows from Lemma 2.28 that $Y \in \mathcal{G}$ is a hyperplane of X if and only if $f(Y)$ is a hyperplane of $f(X)$. Denote by f_X the restriction of f to $\mathcal{G}^1(X)$. This is a bijection to $\mathcal{G}^1(f(X))$. Every apartment $\mathcal{A} \subset \mathcal{G}^1(X)$ can be extended to an apartment $\mathcal{A}' \subset \mathcal{G}$ and $f_X(\mathcal{A})$ is the intersection of the apartment $f(\mathcal{A}') \subset \mathcal{G}$ with $\mathcal{G}^1(f(X))$. This intersection is an apartment of $\mathcal{G}^1(f(X))$. So, f_X transfers apartments to apartments. The same arguments show that the inverse map also sends apartments to apartments. As in the proof of Theorem 2.24 for $\alpha = 1$, we establish that f_X is induced by a semilinear isomorphism

$$L_X : X \to f(X).$$

This semilinear isomorphism is unique up to a scalar multiple.

Consider any $Y \in \mathcal{G}$ contained in X. Lemma 2.28 shows that Y is contained in $Z \in \mathcal{G}^1(X)$ if and only if $f(Y)$ is contained in $f(Z) = L_X(Z)$. This implies that $f(Y) = L_X(Y)$. By the same arguments, we have $f_Y(H) = f(H) = L_X(H)$ for every $H \in \mathcal{G}^1(Y)$. Therefore, L_Y is a scalar multiple of the restriction of L_X to Y.

For every $X \in \mathcal{G}$ we denote by h_X the bijection of $\mathcal{G}_1(X)$ to $\mathcal{G}_1(f(X))$ induced by L_X. If $Y \in \mathcal{G}$ is contained in X, then the restriction of h_X to $\mathcal{G}_1(Y)$ coincides with h_Y.

Also, if $X, Y \in \mathcal{G}$ have a non-zero intersection, then $h_X(P) = h_Y(P)$ for every 1-dimensional subspace $P \subset X \cap Y$. Indeed, if $X \cap Y$ is an element of $\mathcal{G}$, then

$$h_X(P) = h_{X \cap Y}(P) = h_Y(P).$$

In the case when $X \cap Y$ does not belong to $\mathcal{G}$, there exists $Z \in \mathcal{G}$ containing $X \cap Y$ and intersecting both X, Y in some elements of $\mathcal{G}$. Then

$$h_X(P) = h_Z(P) = h_Y(P).$$

Therefore, there is a bijective transformation of $\mathcal{G}_1(V)$ whose restriction to every $\mathcal{G}_1(X)$, $X \in \mathcal{G}$ coincides with h_X. This is an automorphism of the projective space Π_V, i.e. it is induced by a semilinear automorphism of V. This semilinear automorphism induces f. □

Suppose that $\mathcal{G}$ is formed by subspaces whose dimension and codimension both are infinite. Let C be a connected component of the Grassmann graph $\Gamma(\mathcal{G})$. Every non-empty intersection of C and an apartment of $\mathcal{G}$ will be called an *apartment* of C. If $\mathcal{A}$ is an apartment of $\mathcal{G}$ and X belongs to $\mathcal{A} \cap C$, then $\mathcal{A} \cap C$ consists of all $Y \in \mathcal{A}$ satisfying

$$\dim X/(X \cap Y) = \dim Y/(X \cap Y) < \infty.$$

For any two elements of C there is an apartment of C containing them (we take an apartment of $\mathcal{G}$ containing these elements and the intersection of this apartment with C is as required).

Theorem 2.29 (Pankov [47])   *Suppose that V is infinite-dimensional. Let C and C' be connected components of the Grassmann graph associated to a Grassmannian formed by subspaces of V whose dimension and codimension both are infinite. Let also f be a bijection of C to C' such that f and f^{-1} send apartments to apartments. Then f can be uniquely extended to an automorphism of the lattice $\mathcal{L}(V)$.*

Proof   Let $\mathcal{A}_c$ be an apartment of C. Then $\mathcal{A}_c = \mathcal{A} \cap C$, where $\mathcal{A}$ is an apartment of the associated Grassmannian $\mathcal{G}$. A subset of $\mathcal{A}_c$ is called *inexact* if there is an apartment of C distinct from $\mathcal{A}_c$ and containing this subset. A subset of $\mathcal{A}_c$ is inexact if and only if it is the intersection of C with an inexact subset of $\mathcal{A}$; moreover, the maximal inexact subsets of $\mathcal{A}_c$ are just the intersections of C with maximal inexact subsets of $\mathcal{A}$. We say that $X \subset \mathcal{A}_c$ is a *complementary* subset if $\mathcal{A}_c \setminus X$ is a maximal inexact subset of $\mathcal{A}_c$. Complementary subsets of $\mathcal{A}_c$ can be characterized as the intersections of C with complementary subsets of $\mathcal{A}$.

A bijection g from an apartment of C to an apartment of C' is called *special* if g and g^{-1} transfer inexact subsets to inexact subsets. For every special bijection one of the following possibilities is realized:

(1) The intersections of C with simple subsets of first and second type go to the intersections of C' with simple subsets of first and second type, respectively.

(2) The intersections of C with simple subsets of first type go to the intersections of C' with simple subsets of second type, and the intersections of C with simple subsets of second type correspond to the intersections of C' with simple subsets of first type. In this case, we have $\mathcal{G} = \mathcal{G}_\alpha(V)$, where $\dim V = \alpha$.

The proof is similar to the proof of Lemma 2.26.

The restrictions of f to apartments of C are special bijections. As in the proof of Lemma 2.27, we show that f is an isomorphism between the restrictions of the Grassmann graph to C and C'. The case (2) is not realized for the restrictions of f to apartments of C. This means that f sends stars to stars and tops to tops, i.e. f can be uniquely extended to a lattice isomorphism $g : C_\pm \to C'_\pm$ (Theorem 2.19).

If $\mathcal{A}$ is an apartment of $\mathcal{G}$ and $\mathcal{A} \cap C$ is an apartment of C, then $f(\mathcal{A} \cap C)$ is an apartment of C' and

$$f(\mathcal{A} \cap C) = \mathcal{A}' \cap C',$$

where $\mathcal{A}'$ is an apartment of $\mathcal{G}$. Since $g : C_\pm \to C'_\pm$ is a lattice isomorphism, we have

$$g(\mathcal{A} \cap C_\pm) = \mathcal{A}' \cap C'_\pm.$$

Therefore, g sends the intersections of $C_\pm$ with apartments of $\mathcal{G}$ to the intersections of $C'_\pm$ with apartments of $\mathcal{G}$ and the same holds for g^{-1}.

For every $X \in C_\pm$ we denote by g_X the restriction of g to $\mathcal{G}^1(X)$. This is a bijection to $\mathcal{G}^1(g(X))$ such that g_X and g_X^{-1} send apartments to apartments. This implies the existence of a semilinear isomorphism $L_X : X \to g(X)$ such that $g(Y) = L_X(Y)$ for every $Y \in \mathcal{G}^1(X)$. The remaining part of the proof is similar to the proof of Theorem 2.24 for the case when α is infinite.   □

Remark 2.30   There are more general results concerning arbitrary (not necessarily bijective) apartments preserving transformations of Grassmannians of finite-dimensional vector spaces; see [45, Chapter 5].

3

Lattices of Closed Subspaces

The main objects of this chapter are the lattices formed by closed subspaces of infinite-dimensional complex normed spaces. Our first result is the following analogue of the Fundamental Theorem of Projective Geometry: all isomorphisms of such lattices are induced by linear or conjugate-linear homeomorphisms between the corresponding normed spaces (for the finite-dimensional case this fails). This statement is closely connected to the remarkable Kakutani–Mackey theorem [31] which states that every orthomodular lattice consisting of all closed subspaces of an infinite-dimensional complex Banach space is the orthomodular lattice associated to an infinite-dimensional complex Hilbert space.

At the end, we consider the partially ordered set formed by all closed subspaces of a complex Hilbert space whose dimension and codimension both are infinite. It will be shown that every isomorphism between such partially ordered sets can be uniquely extended to an isomorphism of the lattices of closed subspaces. Using the same arguments, we obtain an analogue of Chow's theorem for connected components of the Grassmamm graphs associated to infinite-dimensional Hilbert spaces (note that for Grassmannians of infinite-dimensional vector spaces there is no result of such kind).

3.1 Linear and Conjugate-Linear Operators

We will work with semilinear maps of complex normed spaces. For this reason, we need some information on endomorphisms of the field of complex numbers. The automorphism group of this field contains the conjugate map $a \to \bar{a}$ and infinitely many other elements.

Example 3.1 Using Zorn's lemma and [34, Chapter V, Theorem 2.8], we can show that every automorphism of a field can be extended to an automorphism of any algebraically closed extension of this field (see, for example, [45, Section 1.1]). The field $\mathbb{Q}(\sqrt{p})$, where p is a prime number, is contained in the algebraically closed field $\mathbb{C}$. Consider the automorphism of $\mathbb{Q}(\sqrt{p})$ sending every $v + w\sqrt{p}$ to $v - w\sqrt{p}$ and extend it to an automorphism of $\mathbb{C}$. Any such extension is not identity on $\mathbb{R}$, which implies that it is different from the conjugate map.

Lemma 3.2 *Every continuous endomorphism of the field $\mathbb{C}$ is identity or the conjugate map.*

Proof If σ is an endomorphism of $\mathbb{C}$, then the restriction of σ to $\mathbb{Q}$ is identity. Therefore, the restriction of σ to $\mathbb{R}$ is identity if σ is continuous. It is clear that $\sigma(\mathbf{i}) = \pm\mathbf{i}$ and we get the claim. □

A *complex normed vector space* is a complex vector space N together with a real-valued norm function $x \rightarrow \|x\|$ satisfying the following conditions:

- $\|x\| \geq 0$ for every $x \in N$ and $\|x\| = 0$ only in the case when $x = 0$,
- $\|ax\| = |a| \cdot \|x\|$ for all $x \in N$ and $a \in \mathbb{C}$,
- $\|x + y\| \leq \|x\| + \|y\|$ for all $x, y \in N$.

A normed vector space can be considered as a metric space, where the distance between any two vectors x and y is equal to $\|x - y\|$. In the case when this metric space is complete, the normed space is called a *Banach space*.

Let N and N' be complex normed spaces. A semilinear map $L : N \rightarrow N'$ is *bounded* if there is a non-negative real number a such that

$$\|L(x)\| \leq a\|x\|$$

for all vectors $x \in N$. The smallest number a satisfying this condition is called the *norm* of L and denoted by $\|L\|$.

Proposition 3.3 *For every bounded semilinear map between complex normed spaces the associated endomorphism of the field $\mathbb{C}$ is identity or the conjugate map.*

This statement is a simple consequence of the following lemma (which will be also exploited in the next section).

Lemma 3.4 *If σ is an endomorphism of the field $\mathbb{C}$ such that for every sequence of complex numbers $\{a_n\}_{n\in\mathbb{N}}$ converging to zero the sequence $\{\sigma(a_n)\}_{n\in\mathbb{N}}$ is bounded, then σ is identity or the conjugate map.*

Proof By Lemma 3.2, we need to show that σ is continuous. Since σ is additive, it is sufficient to establish that σ is continuous in zero.

Suppose that a sequence $\{a_n\}_{n\in\mathbb{N}}$ converges to zero and the same fails for the sequence $\{\sigma(a_n)\}_{n\in\mathbb{N}}$. Then $\{a_n\}_{n\in\mathbb{N}}$ contains a subsequence $\{a'_n\}_{n\in\mathbb{N}}$ such that the inequality $|\sigma(a'_n)| > a$ holds for a certain real number $a > 0$ and all natural n. In the sequence $\{a'_n\}_{n\in\mathbb{N}}$, we choose a subsequence $\{a''_n\}_{n\in\mathbb{N}}$ satisfying $na''_n \to 0$. Since σ is an endomorphism of $\mathbb{C}$, we have $\sigma(n) = n$ for every natural n. Then

$$|\sigma(na''_n)| = n|\sigma(a''_n)| > na$$

and the sequence $\{\sigma(na''_n)\}_{n\in\mathbb{N}}$ is unbounded, which contradicts our assumption. $\square$

Remark 3.5 For a semilinear map $L : N \to N'$ the following two conditions are equivalent:

- L is continuous,
- L is bounded.

This is well known if L is linear, but we need some explanations in the general case. If L is continuous, then the associated endomorphism of $\mathbb{C}$ is continuous, i.e. it is identity or the conjugate map, and for every real $\varepsilon > 0$ there is a real $\delta > 0$ such that $\|L(x)\| < \varepsilon$ if $\|x\| = \delta$; since the field endomorphism associated to L preserves each real number, we have

$$\|L(x)\| < \varepsilon\delta^{-1}\|x\|$$

for every $x \in N$ and L is bounded. Conversely, suppose that L is bounded. Then it sends every vector sequence converging to zero to a bounded sequence. As in the proof of Lemma 3.4, we show that $L(x_n) \to 0$ for every vector sequence $x_n \to 0$, i.e. L is continuous in zero. Then L is continuous by additivity.

Linear maps of normed spaces are called *linear operators*. A semilinear map between complex normed spaces is said to be a *conjugate-linear operator* if the associated endomorphism of $\mathbb{C}$ is the conjugate map.

Example 3.6 Let H be a complex Hilbert space and let $B = \{e_i\}_{i\in I}$ be an orthonormal basis of H. There is a unique conjugate-linear operator C_B which leaves fixed every vector from this basis. If J is a countable or finite subset of I and $x = \sum_{j\in J} a_j e_j$, then

$$C_B(x) = \sum_{j\in J} \overline{a}_j e_j.$$

Since $\|C_B(x)\| = \|x\|$ for every vector $x \in H$, the operator C_B is bounded.

Denote by $\mathcal{L}(N)$ and $\mathcal{L}(N')$ the sets of all closed subspaces of N and N', respectively. The partially ordered sets $(\mathcal{L}(N), \subset)$ and $(\mathcal{L}(N'), \subset)$ are bounded lattices (such lattices were considered in Section 1.3 for Hilbert spaces). Every linear or conjugate-linear homeomorphism $A : N \to N'$ induces an isomorphism between these lattices and each non-zero scalar multiple of A defines the same lattice isomorphism. In the case when N and N' are Banach spaces, every invertible bounded linear or conjugate linear operator $A : N \to N'$ is a homeomorphism (if A is linear, then this follows easily from the Open Map Theorem, see [55, Corollary 2.12]; readers can check that the Open Map Theorem [55, Theorem 2.11] holds for the conjugate-linear maps).

Let H and H' be complex Hilbert spaces. Recall that for every bounded linear operator $A : H \to H'$ the adjoint linear operator $A^* : H' \to H$ satisfies

$$\langle A(x), y \rangle = \langle x, A^*(y) \rangle$$

for all $x \in H$ and $y \in H'$. Now, we consider the case when $A : H \to H'$ is a bounded conjugate-linear operator. For every vector $y \in H'$ the map

$$x \to \overline{\langle A(x), y \rangle}$$

is a linear functional on H and there is a unique vector $A^*(y) \in H$ such that

$$\overline{\langle A(x), y \rangle} = \langle x, A^*(y) \rangle$$

for all vectors $x \in H$. So, we get a conjugate-linear operator $A^* : H' \to H$ which will be called *adjoint* to A. As in the linear case, the operator A^* is bounded and $\|A^*\| = \|A\|$.

For every linear or conjugate-linear bounded operator $A : H \to H'$ we have $A^{**} = A$. The kernel of A^* is the orthogonal complement of the image of A. Therefore, A^* is invertible if and only if A is invertible. In this case, the operators $(A^{-1})^*$ and $(A^*)^{-1}$ are coincident. For every scalar $a \in \mathbb{C}$ we have $(aA)^* = \overline{a}A^*$ if A is linear, and $(aA)^* = aA^*$ if it is conjugate-linear.

There is the following relation between adjoint operators and lattice isomorphisms.

Proposition 3.7 *For every invertible bounded linear or conjugate-linear operator $A : H \to H'$ the map of $\mathcal{L}(H)$ to $\mathcal{L}(H')$ defined as*

$$X \to A(X^\perp)^\perp$$

for every $X \in \mathcal{L}(H)$ is the lattice isomorphism induced by the operator $(A^*)^{-1}$.

Proof If $x, y \in H$, then $\langle y, x \rangle = \langle A^{-1}A(y), x \rangle$ is equal to

$$\langle A(y), (A^{-1})^*(x) \rangle \quad \text{or} \quad \overline{\langle A(y), (A^{-1})^*(x) \rangle}.$$

In other words, y is orthogonal to x if and only if $A(y)$ is orthogonal to $(A^{-1})^*(x)$. This implies that

$$A(X^\perp)^\perp = (A^{-1})^*(X) = (A^*)^{-1}(X)$$

for every closed subspace $X \subset H$. $\qquad\qquad\qquad\qquad\qquad\qquad\qquad\qquad$ □

3.2 Lattice Isomorphisms

We prove the following analogue of the Fundamental Theorem of Projective Geometry for the lattices of closed subspaces of infinite-dimensional complex normed spaces.

Theorem 3.8 (Mackey [35] and Kakutani and Mackey [31]) *Let N and N' be infinite-dimensional complex normed spaces. Then every isomorphism between the lattices $\mathcal{L}(N)$ and $\mathcal{L}(N')$ is induced by a linear or conjugate-linear homeomorphism $A : N \to N'$ and any other operator inducing this lattice isomorphism is a scalar multiple of A.*

Remark 3.9 For finite-dimensional complex normed spaces this statement fails. Indeed, if N is a complex normed space of finite dimension, then $\mathcal{L}(N)$ consists of all subspaces of N (since every finite-dimensional subspace is closed) and some automorphisms of $\mathcal{L}(N)$ are induced by unbounded semilinear automorphisms of N, i.e. semilinear automorphisms associated to non-continuous automorphisms of the field $\mathbb{C}$.

Let f be an isomorphism between the lattices $\mathcal{L}(N)$ and $\mathcal{L}(N')$, where N and N' are complex normed spaces. Then $f(\mathcal{G}_1(N)) = \mathcal{G}_1(N')$ and the restriction of f to $\mathcal{G}_1(N)$ is an isomorphism between the projective spaces Π_N and $\Pi_{N'}$, i.e. this restriction is induced by a semilinear isomorphism $L : N \to N'$. It is easy to see that $f(X) = L(X)$ for every $X \in \mathcal{L}(N)$.

Now, we suppose that our normed spaces are infinite-dimensional and show that L is linear or conjugate-linear. After that we establish that L is a homeomorphism.

Lemma 3.10 *If N is infinite-dimensional, then there exist a linearly independent set of vectors $\{x_n\}_{n\in\mathbb{N}}$ and a sequence of bounded linear functionals $\{v_n\}_{n\in\mathbb{N}}$ on N such that*

$$v_i(x_j) = \delta_{ij}. \qquad\qquad (3.1)$$

Proof We take any closed hyperplane $H \subset N$ and a vector $x_1 \notin H$. Consider the bounded linear functional v_1 such that $v_1(tx_1 + y) = t$ for every vector

$y \in H$ and every scalar t. Suppose that (3.1) holds for linearly independent vectors $x_1, \ldots, x_n$ and bounded linear functionals $v_1, \ldots, v_n$. Since N is infinite-dimensional, there are a bounded linear functional v_{n+1} whose kernel contains all x_i and a vector x'_{n+1} such that $v_{n+1}(x'_{n+1}) = 1$ [55, Theorem 3.5]. We define

$$x_{n+1} = x'_{n+1} - \sum_{i=1}^{n} v_i(x'_{n+1})x_i.$$

Then $v_{n+1}(x_{n+1}) = 1$ and $v_i(x_{n+1}) = 0$ for all $i \le n$.   □

Lemma 3.11   *If N is infinite-dimensional, then it contains a linearly independent subset $\{x_n\}_{n \in \mathbb{N}}$ satisfying the following condition: for every bounded sequence of scalars $\{a_n\}_{n \in \mathbb{N}}$ there is a bounded linear functional v on N such that*

$$v(x_n) = a_n$$

for every natural n.

Proof   Let $\{x_n\}_{n \in \mathbb{N}}$ and $\{v_n\}_{n \in \mathbb{N}}$ be as in the previous lemma. We can assume that

$$\|v_n\| = 1/2^n   \text{for all } n \in \mathbb{N}$$

(for every natural n there is a scalar b_n such that $\|b_n v_n\| = 1/2^n$ and we take a scalar multiple x'_n of x_n satisfying $b_n v_n(x'_n) = 1$).

Let $\{a_n\}_{n \in \mathbb{N}}$ be a bounded sequence of scalars and $a = \sup |a_i|$. Denote by X the subspace formed by all linear combinations of vectors from the sequence $\{x_n\}_{n \in \mathbb{N}}$. For every vector

$$x = v_1(x)x_1 + \cdots + v_n(x)x_n \in X$$

we define

$$v(x) := v_1(x)a_1 + \cdots + v_n(x)a_n.$$

Then

$$|v(x)| \le |a_1| \cdot \|v_1\| \cdot \|x\| + \cdots + |a_n| \cdot \|v_n\| \cdot \|x\|$$

$$\le a\|x\|(1/2 + \cdots + 1/2^n) < a\|x\|.$$

This means that v is a bounded linear functional on X. By [55, Theorem 3.6], v can be extended to a bounded linear functional of N.   □

The following lemma is crucial in our proof.

Lemma 3.12 (Kakutani and Mackey [31]) *Suppose that N and N′ are infinite-dimensional. If a semilinear isomorphism L : N → N′ sends closed hyperplanes to closed hyperplanes, then L is linear or conjugate-linear.*

Proof Let σ be the automorphism of $\mathbb{C}$ associated to L. Let also $\{x_n\}_{n\in\mathbb{N}}$ be a subset of N with the property described in the previous lemma. By Lemma 3.4, we need to show that for every sequence of complex numbers $\{a_n\}_{n\in\mathbb{N}}$ converging to zero the sequence $\{\sigma(a_n)\}_{n\in\mathbb{N}}$ is bounded.

If the latter sequence is unbounded, then $\{a_n\}_{n\in\mathbb{N}}$ contains a subsequence $\{b_n\}_{n\in\mathbb{N}}$ such that

$$|\sigma(b_n)| \geq n\|L(x_n)\| \tag{3.2}$$

for all natural n. By Lemma 3.11, there is a bounded linear functional v on N satisfying $v(x_n) = b_n$ for every n. We take any vector $z \in N$ such that $v(z) = 1$. Then $x_n = y_n + b_n z$, where $y_n \in \text{Ker}\, v$. We have

$$L(x_n)/\sigma(b_n) = L(y_n/b_n) + L(z)$$

and (3.2) implies that

$$L(x_n)/\sigma(b_n) \to 0 \quad \text{and} \quad L(-y_n/b_n) \to L(z).$$

The latter means that $L(z)$ belongs to the closure of $L(\text{Ker}\, v)$. On the other hand, $\text{Ker}\, v$ is a closed hyperplane and the same holds for $L(\text{Ker}\, v)$ by our assumption. Therefore, $L(z)$ belongs to $L(\text{Ker}\, v)$. Since L is bijective, we get $z \in \text{Ker}\, v$, which contradicts $v(z) = 1$. □

Proof of Theorem 3.8 It was noted above that every isomorphism between the lattices $\mathcal{L}(N)$ and $\mathcal{L}(N')$ is induced by a semilinear isomorphism $L : N \to N'$. Then L and L^{-1} send closed hyperplanes to closed hyperplanes and Lemma 3.12 implies that L is linear or conjugate-linear.

Let v be a non-zero bounded linear functional on N'. Then $\text{Ker}\, v$ is a closed hyperplane of N' and $S = L^{-1}(\text{Ker}\, v)$ is a closed hyperplane of N. For every closed hyperplane of N there is a bounded linear functional whose kernel coincides with this hyperplane. Consider a bounded linear functional w on N such that $\text{Ker}\, w = S$ and fix $z \in N$ satisfying $w(z) = 1$. Every vector $x \in N$ can be presented as the sum $x = y + w(x)z$, where $y \in S$. Then we have

$$v(L(x)) = v(L(y)) + v(L(w(x)z)) = v(L(w(x)z))$$

(since $L(y) \in \text{Ker}\, v$) and

$$v(L(x)) = w(x)v(L(z)) \quad \text{or} \quad v(L(x)) = \overline{w(x)}v(L(z)) \tag{3.3}$$

(if L is linear or conjugate-linear, respectively).

Let X be a bounded subset of N. Then $w(X)$ is a bounded subset of $\mathbb{C}$ and (3.3) implies that the same holds for $v(L(X))$. Since v is taken arbitrarily, the set $v(L(X))$ is bounded for every bounded linear functional v on N', i.e. $L(X)$ is weakly bounded in N'. In a normed space, every weakly bounded subset is bounded [55, Theorem 3.18]. So, L transfers bounded subsets to bounded subsets, which means that L is bounded.

Similarly, we show that L^{-1} is bounded. $\quad\square$

Remark 3.13 Theorem 3.8 was first proved by Mackey [35] for the lattices of closed subspaces of infinite-dimensional real normed spaces. Lemma 3.12 was obtained in [31]; it shows that the arguments given in [35] work for the complex case. See also [22].

Corollary 3.14 *If H and H' are infinite-dimensional complex Hilbert spaces, then for every anti-isomorphism f of $\mathcal{L}(H)$ to $\mathcal{L}(H')$ there is an invertible bounded linear or conjugate-linear operator $A : H \to H'$ such that*

$$f(X) = A(X)^{\perp}$$

for every $X \in \mathcal{L}(H)$ and any other operator inducing this anti-isomorphism is a scalar multiple of A.

Proof We apply Theorem 3.8 to the composition of f and the orthocomplementation map. $\quad\square$

The algebra of all bounded linear operators on a normed space N is denoted by $\mathcal{B}(N)$. All ring isomorphisms between the algebras of linear operators on real normed spaces were described by Eidelheit [20]. Following [22] we combine the methods of [20] together with the arguments used to prove Theorem 3.8 and get a simple proof of the complex version of Eidelheit's theorem.

Theorem 3.15 (Arnold [2]) *Let N and N' be infinite-dimensional complex normed spaces. For every ring isomorphism $f : \mathcal{B}(N) \to \mathcal{B}(N')$ there is an invertible bounded linear or conjugate-linear operator $L : N \to N'$ such that*

$$f(A) = LAL^{-1} \tag{3.4}$$

for every $A \in \mathcal{B}(N)$.

Proof Let P be a rank one idempotent of $\mathcal{B}(N)$ (see Section 1.3). Then $f(P)$ is an idempotent of $\mathcal{B}(N')$. We fix non-zero vectors x_0 and x_0' belonging to the images of P and $f(P)$, respectively. For any vector $x \in N$ there is $A \in \mathcal{B}(N)$ satisfying $x = A(x_0)$ and we set

$$L(x) = f(A)[x_0'].$$

If $B \in \mathcal{B}(N)$ and $B(x_0) = x$, then AP and BP are rank one operators sending x_0 to x. We have $AP = BP$, which implies that

$$f(A)f(P) = f(B)f(P) \quad \text{and} \quad f(A)[x_0'] = f(B)[x_0'].$$

Therefore, the map $L : N \to N'$ is well defined. It is easy to see that this map is additive and bijective.

If $A, B \in \mathcal{B}(N)$ and $B(x_0) = x$, then

$$LA(x) = LAB(x_0) = f(AB)[x_0'] = f(A)f(B)[x_0'] = f(A)L(x),$$

i.e. $LA(x) - f(A)L(x)$ for every $x \in N$, which means that the equality (3.4) holds for all $A \in \mathcal{B}(N)$.

The centres of $\mathcal{B}(N)$ and $\mathcal{B}(N')$ are formed by all scalar multiples of the identity transformations of N and N', respectively. We have

$$f(a \operatorname{Id}_N) = \sigma(a)\operatorname{Id}_{N'},$$

where σ is an automorphism of the field $\mathbb{C}$. Since

$$f(aA) = f(a \operatorname{Id}_N)f(A) = \sigma(a)f(A)$$

for every $a \in \mathbb{C}$, the map L is σ-linear.

For every closed hyperplane $S \subset N$ there is non-zero $A \in \mathcal{B}(N)$ whose kernel coincides with S. The kernel of $f(A)$ contains $L(S)$. If the hyperplane $L(S)$ is not closed, then its closure coincides with N' and $f(A) = 0$ by continuity. The latter is impossible and $L(S)$ is closed. Similarly, we show that L^{-1} sends closed hyperplanes to closed hyperplanes. Then L is linear or conjugate-linear by Lemma 3.12. As in the proof of Theorem 3.8, we establish that L is bounded. □

3.3 Kakutani–Mackey Theorem

In this section, we consider the lattice $\mathcal{L}(B)$ formed by closed subspaces of an infinite-dimensional complex Banach space B. We show that this lattice is orthomodular only in the case when it is the lattice of closed subspaces of a complex Hilbert space.

Theorem 3.16 (Kakutani and Mackey [31]) *Let B be an infinite-dimensional complex Banach space. Suppose that there is a bijective transformation $X \to X^\perp$ of the lattice $\mathcal{L}(B)$ satisfying the following conditions for any $X, Y \in \mathcal{L}(B)$:*

(1) *the inclusion $X \subset Y$ implies that $Y^\perp \subset X^\perp$,*
(2) $X^{\perp\perp} = X$,

(3) $X \cap X^{\perp} = 0.$

Then there is an inner product $B \times B \to \mathbb{C}$ such that the following assertions are fulfilled:

- *The vector space B together with this inner product is a complex Hilbert space.*
- *The identity transformation of B is an invertible bounded linear operator of the Banach space to the Hilbert space, i.e. a subspace of B is closed in the Banach space if and only if it is closed in the Hilbert space[1].*
- *For every $X \in \mathcal{L}(B)$ the subspace $X^{\perp}$ is the orthogonal complement of X with respect to the inner product.*

Proof Consider the Banach space B^* formed by all bounded linear functionals on B [55, Theorem 4.1]. For every closed subspace $X \subset B$ we denote by X^0 the annihilator of $X^{\perp}$ in N^*, i.e. the set of all bounded linear functionals $v : B \to \mathbb{C}$ satisfying $v(X^{\perp}) = 0$. This is a closed subspace of B^*. If P is a 1-dimensional subspace of B, then (1) shows that $P^{\perp}$ is a closed hyperplane of B and P^0 is a 1-dimensional subspace of B^*. Consider the map of $\mathcal{G}_1(B)$ to $\mathcal{G}_1(B^*)$ which sends every 1-dimensional subspace P to P^0. It follows from (1) that this is an isomorphism between the projective spaces Π_B and Π_{B^*}, i.e. there is a semilinear isomorphism $L : B \to B^*$ such that $L(P) = P^0$ for every $P \in \mathcal{G}_1(B)$.

Let H be a closed hyperplane of B. Then $H = P^{\perp}$ for a certain $P \in \mathcal{G}_1(B)$. By (1) and (2), $Q \in \mathcal{G}_1(B)$ is contained in H if and only if $P \subset Q^{\perp}$; in other words, the kernel of every $v \in L(Q)$ contains P. Then $L(H)$ consists of all $v \in B^*$ satisfying $v(P) = 0$ (since $Q \in \mathcal{G}_1(B)$ is contained in H if and only if $L(Q)$ is contained in $L(H)$). Therefore, $L(H)$ is a closed hyperplane of B^*.

So, L sends closed hyperplanes to closed hyperplanes, and by Lemma 3.12, L is linear or conjugate-linear. As in the proof of Theorem 3.8, we show that the operator $L^{-1} : B^* \to B$ is bounded. This means that L is bounded (it was noted above that the Open Map Theorem [55, Theorem 2.11] holds also for conjugate-linear maps).

Suppose that L is linear. Let x and y be linearly independent vectors of B. We set $l = L(x)$ and $s = L(y)$. Then

$$L(x + ay) = l + as.$$

By (3), we have $X \cap X^{\perp} = 0$ for every $X \in \mathcal{L}(B)$. This implies that each of the

---

[1] We cannot state that the norm related to the inner product coincides with the primordial norm, but these norms define the same topology on B.

scalars

$$l(x), \ s(y), \ (l + as)(x + ay)$$

is non-zero. On the other hand,

$$(l + as)(x + ay) = l(x) + a(l(y) + s(x)) + a^2 s(y)$$

and the equation

$$l(x) + a(l(y) + s(x)) + a^2 s(y) = 0$$

has a solution for a. We get a contradiction. Therefore, L is conjugate-linear.

For all vectors $x, y \in B$ we set

$$\langle x, y \rangle = l(x), \quad \text{where} \ \ l = L(y).$$

Then the product $\langle \cdot, \cdot \rangle$ is linear in the first variable and conjugate-linear in the second. The condition (3) guarantees that $\langle x, x \rangle$ is non-zero for every non-zero vector $x \in B$. Since L can be replaced by any non-zero scalar multiple, we assume that for a certain vector $x_0 \in B$ the scalar $\langle x_0, x_0 \rangle$ is a positive real number.

It follows from (1) that for any two vectors $x, y \in B$ we have $\langle x, y \rangle = 0$ if and only if $\langle y, x \rangle = 0$. Suppose that $\langle x, y \rangle$ is non-zero. We choose non-zero scalars $a, b \in \mathbb{C}$ such that

$$a\langle x, x \rangle + \langle x, y \rangle = 0 = b\langle y, y \rangle + \langle x, y \rangle. \tag{3.5}$$

Then $\langle x, \bar{a}x + y \rangle = 0$, which implies that $\overline{\langle \bar{a}x + y, x \rangle} = 0$ and $\overline{\langle \bar{a}x + y, x \rangle} = 0$, i.e.

$$a\overline{\langle x, x \rangle} + \overline{\langle y, x \rangle} = 0. \tag{3.6}$$

Similarly, we obtain that

$$b\overline{\langle y, y \rangle} + \overline{\langle y, x \rangle} = 0. \tag{3.7}$$

Using (3.5), (3.6) and (3.7), we establish that

$$\overline{\langle x, x \rangle} : \langle x, x \rangle = \overline{\langle y, y \rangle} : \langle y, y \rangle.$$

In other words, for any two vectors $x, y \in B$ satisfying $\langle x, y \rangle \neq 0$ the scalar $\langle x, x \rangle$ is real if and only if $\langle y, y \rangle$ is real. Recall that there is non-zero $x_0 \in B$ such that $\langle x_0, x_0 \rangle$ is real. Then $\langle x, x \rangle$ is real if $\langle x_0, x \rangle$ is non-zero. In the case when $\langle x_0, x \rangle = 0$, we take any vector $y \in B$ such that $\langle x_0, y \rangle$ and $\langle x, y \rangle$ both are non-zero (for example, $y = x_0 + x$ is as required). So, for every non-zero vector $x \in B$ the scalar $\langle x, x \rangle$ is a non-zero real number. Then (3.5) and (3.6) imply that

$$\langle x, y \rangle = \overline{\langle y, x \rangle}.$$

Now, we show that $\langle x, x \rangle$ is a positive real number for every non-zero vector $x \in B$. If $x = sx_0$ for a certain scalar s, then $\langle x, x \rangle = |s|^2 \langle x_0, x_0 \rangle > 0$. In the case when x and x_0 are linearly independent, we consider the real function

$$h(t) = \langle tx + (1 - t)x_0, tx + (1 - t)x_0 \rangle$$

defined on the segment $[0, 1]$. The function is continuous (because L is bounded) and $h(t)$ is non-zero for every $t \in [0, 1]$. Since $h(0) > 0$, we have always $h(t) > 0$.

So, $\langle \cdot, \cdot \rangle$ is an inner product on B. Using the fact that L is bounded, we show that the identity transformation of B is an invertible bounded linear operator of the Banach space to the normed vector space with respect to the inner product $\langle \cdot, \cdot \rangle$. This guarantees that the norm defined by the inner product is complete.

$\square$

3.4 Extensions of Isomorphisms

For an infinite-dimensional complex Hilbert space H we denote by $\mathcal{G}_\infty(H)$ the set of all closed subspaces of H whose dimension and codimension both are infinite. Note that the partially ordered set $(\mathcal{G}_\infty(H), \subset)$ is not a lattice.

Theorem 3.17 (Pankov [42, 43])   *Let H and H' be infinite-dimensional complex Hilbert spaces. Then every isomorphism between the partially ordered sets $(\mathcal{G}_\infty(H), \subset)$ and $(\mathcal{G}_\infty(H'), \subset)$ can be uniquely extended to an isomorphism between the lattices $\mathcal{L}(H)$ and $\mathcal{L}(H')$.*

Since the orthocomplementation map sends $\mathcal{G}_\infty(H)$ to itself, Theorem 3.17 implies the following.

Corollary 3.18   *If H and H' are as in Theorem 3.17, then every anti-isomorphism of $(\mathcal{G}_\infty(H), \subset)$ to $(\mathcal{G}_\infty(H'), \subset)$ can be uniquely extended to an anti-isomorphism between the lattices $\mathcal{L}(H)$ and $\mathcal{L}(H')$.*

Let f be an isomorphism of the partially ordered set $\mathcal{G}_\infty(H)$ to the partially ordered set $\mathcal{G}_\infty(H')$.

Lemma 3.19   *For every $X \in \mathcal{G}_\infty(H)$ there is an invertible bounded linear or conjugate-linear operator $A_X : X \to f(X)$ such that*

$$f(Y) = A_X(Y)$$

for every $Y \in \mathcal{G}_\infty(H)$ contained in X.

Proof Let X be the set of all elements of $\mathcal{G}_\infty(H)$ contained in X. Then $f(X)$ consists of all elements of $\mathcal{G}_\infty(H')$ contained in $X' = f(X)$. We consider the closed subspaces X and X' as Hilbert spaces and write $Y^\perp$ and $Y'^\perp$ for the orthogonal complements of $Y \subset X$ and $Y' \subset X'$ in these Hilbert spaces. Let $\mathcal{Y}$ and $\mathcal{Y}'$ be the sets formed by all closed subspaces of infinite codimension in X and X', respectively. Then

$$Y \in \mathcal{Y} \iff Y^\perp \in X \text{ and } Y' \in \mathcal{Y}' \iff Y'^\perp \in f(X).$$

Denote by g the bijection sending every $Y \in \mathcal{Y}$ to $f(Y^\perp)^\perp$. This is an isomorphism between the partially ordered sets $(\mathcal{Y}, \subset)$ and $(\mathcal{Y}', \subset)$. The restriction of g to $\mathcal{G}_1(X)$ is an isomorphism of Π_X to $\Pi_{X'}$ and there is a semilinear isomorphism $L : X \to X'$ such that $g(Y) = L(Y)$ for every $Y \in \mathcal{G}_1(X)$. The same holds for all $Y \in \mathcal{Y}$ (since g is an isomorphism of partially ordered sets). Also, for every $Y \in \mathcal{Y}$ the lattice $\mathcal{L}(Y)$ is contained in $\mathcal{Y}$ and the restriction of g to this lattice is an isomorphism to the lattice $\mathcal{L}(g(Y))$. Theorem 3.8 implies that L is bounded on any infinite-dimensional subspace $Y \in \mathcal{Y}$. Show that L is bounded.

We take any orthogonal $Y, Z \in \mathcal{Y}$ such that $X = Y \oplus Z$. If a sequence $\{y_i + z_i\}_{i\in\mathbb{N}}$ converges to $y_0 + z_0$ and $y_i \in Y$, $z_i \in Z$ for all $i = 0, 1, \ldots$, then the sequences $\{y_i\}_{i\in\mathbb{N}}$ and $\{z_i\}_{i\in\mathbb{N}}$ converge to y_0 and z_0, respectively. Since L is bounded on Y and Z, the sequences $\{L(y_i)\}_{i\in\mathbb{N}}$ and $\{L(z_i)\}_{i\in\mathbb{N}}$ converge to $L(y_0)$ and $L(z_0)$, respectively. This means that $\{L(y_i + z_i)\}_{i\in\mathbb{N}}$ converges to $L(y_0 + z_0)$. Therefore, L is continuous and, consequently, bounded.

So, $L : X \to X'$ is an invertible bounded linear or conjugate-linear operator. We have

$$f(Y) = L(Y^\perp)^\perp$$

for every $Y \in X$. Proposition 3.7 shows that the operator $A_X = (L^*)^{-1}$ satisfies the required condition. $\qquad\square$

Lemma 3.20 *Let X and Y be elements of $\mathcal{G}_\infty(H)$ satisfying*

$$\dim(X \cap Y) < \infty.$$

Then there exists $Z \in \mathcal{G}_\infty(H)$ such that $X \cap Z$ and $Y \cap Z$ are elements of $\mathcal{G}_\infty(H)$ containing $X \cap Y$.

Proof Let X' and Y' be the orthogonal complements of $X \cap Y$ in X and Y, respectively. The subspaces X' and Y' both are infinite-dimensional and we choose inductively a sequence of mutually orthogonal vectors $\{x_n, x'_n, y_n, y'_n\}_{n\in\mathbb{N}}$ such that

$$x_n, x'_n \in X' \text{ and } y_n, y'_n \in Y'$$

for every $n \in \mathbb{N}$. Denote by Z' the closed subspace spanned by $\{x_n', y_n'\}_{n \in \mathbb{N}}$. The subspace

$$Z = (X \cap Y) + Z'$$

is as required.   □

Proof of Theorem 3.17   For any finite-dimensional subspace $S \subset H$ we take any $X \in \mathcal{G}_\infty(H)$ containing S and set

$$g(S) = A_X(S).$$

We need to show that the definition of $g(S)$ does not depend on the choice of $X \in \mathcal{G}_\infty(H)$.

Suppose that S is contained in $X \in \mathcal{G}_\infty(H)$ and $Y \in \mathcal{G}_\infty(H)$. In the case when $X \cap Y$ is an element of $\mathcal{G}_\infty(H)$, take $X', Y' \in \mathcal{G}_\infty(H)$ contained in $X \cap Y$ and such that $X' \cap Y' = S$. Then

$$A_X(S) = A_X(X') \cap A_X(Y') = f(X') \cap f(Y') = A_Y(X') \cap A_Y(Y') = A_Y(S).$$

If $X \cap Y$ is finite-dimensional, then, by Lemma 3.20, there is $Z \in \mathcal{G}_\infty(H)$ such that $X \cap Z$ and $Y \cap Z$ are elements of $\mathcal{G}_\infty(H)$ containing $X \cap Y$. Applying the above arguments to the pairs X, Z and Y, Z, we establish that

$$A_X(S) = A_Z(S) = A_Y(S).$$

The map g is an isomorphism between the lattices $\mathcal{L}_{\mathrm{fin}}(H)$ and $\mathcal{L}_{\mathrm{fin}}(H')$, hence it is induced by a semilinear isomorphism $A : H \to H'$. For every $X \in \mathcal{G}_\infty(H)$ we have $f(X) = A(X)$ and the restriction of A to X is a scalar multiple of A_X. Since each A_X is bounded and H can be presented as the sum of two orthogonal elements of $\mathcal{G}_\infty(H)$, the operator A is bounded, i.e. it is an invertible bounded linear or conjugate-linear operator.   □

Remark 3.21   The above proof is an essential modification of the proof given in [42, 43].

Remark 3.22   Let H be a complex Hilbert space and let $\mathcal{I}(H)$ be the set of all idempotents of the algebra $\mathcal{B}(H)$. The set $\mathcal{I}(H)$ is partially ordered as follows: for $P, Q \in \mathcal{I}(H)$ we have $P \leq Q$ if

$$\mathrm{Im}(P) \subset \mathrm{Im}(Q) \quad \text{and} \quad \mathrm{Ker}(Q) \subset \mathrm{Ker}(P).$$

Since every closed subspace $X \subset H$ can be identified with the projection P_X, the lattice $\mathcal{L}(H)$ is contained in this partially ordered set. Ovchinikov [41] proved that every automorphism of the partially ordered set $\mathcal{I}(H)$ is of type

$$P \to APA^{-1} \quad \text{or} \quad P \to AP^*A^{-1},$$

where A is an invertible bounded linear or conjugate-linear operator on H. Now, we suppose that H is infinite-dimensional and consider the partially ordered set $\mathcal{I}_\infty(H)$ formed by all idempotents from $\mathcal{I}(H)$ whose image and kernel both are infinite-dimensional. Plevnik [51] showed that every automorphism of this partially ordered set can be uniquely extended to an automorphism of the partially ordered set $\mathcal{I}(H)$.

For an infinite-dimensional complex Hilbert space H we denote by $\Gamma_\infty(H)$ the graph whose vertex set is $\mathcal{G}_\infty(H)$ and whose edges are pairs of adjacent elements $X, Y \in \mathcal{G}_\infty(H)$, i.e. such that $X \cap Y$ is a hyperplane in both X and Y. This graph is not connected. A connected component containing $X \in \mathcal{G}_\infty(H)$ consists of all $Y \in \mathcal{G}_\infty(H)$ satisfying

$$\dim X/(X \cap Y) = \dim Y/(X \cap Y) < \infty.$$

The restriction of every automorphism or anti-automorphism of the lattice $\mathcal{L}(H)$ to $\mathcal{G}_\infty(H)$ is an automorphism of $\Gamma_\infty(H)$, but there are automorphisms of $\Gamma_\infty(H)$ which cannot be extended to automorphisms or anti-automorphisms of $\mathcal{L}(H)$ (a simple modification of Example 2.18).

Let C be a connected component of $\Gamma_\infty(H)$. As in Chapter 2, we denote by $C_\pm$ the set of all $X \in \mathcal{G}_\infty(H)$ such that X is a subspace of finite codimension in a certain element of C or X contains an element of C as a subspace of finite codimension. Note that $(C_\pm, \subset)$ is an unbounded lattice.

Theorem 3.23 *Let C and C' be connected components of the graph $\Gamma_\infty(H)$. If an automorphism of $\Gamma_\infty(H)$ sends C to C', then the restriction of this automorphism to C can be uniquely extended to an isomorphism or anti-isomorphism of $(C_\pm, \subset)$ to $(C'_\pm, \subset)$. Every automorphism or anti-isomorphism between these lattices can be uniquely extended to an automorphism or, respectively, an anti-automorphism of the lattice $\mathcal{L}(H)$.*

The proof of the first part of Theorem 3.23 is similar to the proof of Theorem 2.19. Using arguments from the proof of Theorem 3.17, we establish that every isomorphism of $(C_\pm, \subset)$ to $(C'_\pm, \subset)$ is uniquely extendable to an automorphism of $\mathcal{L}(H)$. The composition of every anti-isomorphism f of $(C_\pm, \subset)$ to $(C'_\pm, \subset)$ and the orthocomplementation is an isomorphism of $(C_\pm, \subset)$ to $(C''_\pm, \subset)$, where C'' is the connected component of the graph $\Gamma_\infty(H)$ formed by the orthogonal complements of elements from $C'_\pm$. This composition can be uniquely extended to an automorphism of $\mathcal{L}(H)$, which implies that f is uniquely extendable to an anti-automorphism of $\mathcal{L}(H)$.

4

Wigner's Theorem and Its Generalizations

Recall that the classic Wigner's theorem [67] states that every bijective transformation of the Grassmannian formed by 1-dimensional subspaces of a complex Hilbert space (the set of pure states) preserving the angles between subspaces (or, equivalently, the transition probability) is induced by a unitary or anti-unitary operator. In the case when the dimension of the Hilbert space is not less than three, this statement follows immediately from Uhlhorn's description of bijective transformations preserving the orthogonality relation in both directions [62]. However, there is a non-bijective version of Wigner's theorem characterizing linear and conjugate-linear isometries and there exist non-bijective transformations which preserve the orthogonality relation in both directions and cannot be obtained from such isometries.

All results mentioned above concern the Grassmannian of 1-dimensional subspaces. Molnár [36, 38] extended Wigner's theorem on Grassmannians formed by finite-dimensional subspaces of complex Hilbert spaces. He proved that all (not necessarily bijective) transformations of such Grassmannians preserving the principal angles between subspaces are induced by linear or conjugate-linear isometries (in the case when the Grassmannian is formed by subspaces whose dimension is equal to the codimension, there are also the compositions of transformations induced by isometries and the orthocomplementation). We show that the latter statement holds for transformations preserving only some types of the principal angles. Next, we present Gehér's generalization of Molnár's theorem which describes trace preserving transformations [25]. We also investigate orthogonality preserving transformations. In particular, we give a new proof of the Győry–Šemrl theorem [27, 59] (an extension of Uhlhorn's result on other Grassmannians) based on Chow's theorem and show that a non-bijective version of this statement holds only for finite-dimensional Hilbert spaces. At the end, following [24] we consider bijective transformations of Grassmannians which preserve the gap metric.

4.1 Linear and Conjugate-Linear Isometries

Let H be a complex Hilbert space. The dimension of H is always assumed to be not less than three (except in Section 4.3).

A semilinear map $A : H \to H$ is an *isometry* if it preserves the norm of vectors, i.e.

$$\|A(x)\| = \|x\|$$

for all $x \in H$. It is clear that such a semilinear map is injective and bounded. The latter implies that it is linear or conjugate-linear. The well-known equalities

$$\mathrm{Re}\langle x, y \rangle = \frac{1}{2}(\|x + y\|^2 - \|x\|^2 - \|y\|^2)$$

and

$$\mathrm{Im}\langle x, y \rangle = \frac{1}{4}(\|x + iy\|^2 - \|x - iy\|^2)$$

show that every linear isometry $A : H \to H$ preserves the inner product, i.e.

$$\langle A(x), A(y) \rangle = \langle x, y \rangle$$

for all $x, y \in H$. If $A : H \to H$ is a conjugate-linear isometry, then we get that

$$\langle A(x), A(y) \rangle = \overline{\langle x, y \rangle} \tag{4.1}$$

for all $x, y \in H$. For each of these cases, the composition A^*A is identity.

An invertible bounded linear operator is called *unitary* if it preserves the inner product. Similarly, we say that an invertible bounded conjugate-linear operator A on H is *anti-unitary* if it satisfies (4.1) for all $x, y \in H$. An invertible bounded operator A is unitary or anti-unitary if and only if $A^* = A^{-1}$.

If H is finite-dimensional, then every semilinear isometry of H to itself is surjective, i.e. it is a unitary or anti-unitary operator. In the case when H is infinite-dimensional, there are non-surjective semilinear isometries of H to itself. If $\{e_i\}_{i \in I}$ is an orthonormal basis of H and $\{e_i'\}_{i \in I}$ is an orthonormal basis of a closed subspace of H, then the correspondence $e_i \to e_i'$ can be uniquely extended to a non-surjective linear isometry and there is also the unique extension to a non-surjective conjugate-linear isometry.

Example 4.1 For every orthonormal basis B of H the operator C_B considered in Example 3.6 is anti-unitary. Every conjugate-linear isometry $A' : H \to H$ sends an orthonormal basis B of H to an orthonormal basis of $A'(H)$ which can be extended to an orthonormal basis B' of H (such an extension is not unique). Let A be the linear isometry such that $A(x) = A'(x)$ for every $x \in B$. Then $A' = AC_B = C_{B'}A$.

Proposition 4.2   *If an injective semilinear transformation of H sends orthogonal vectors to orthogonal vectors, then it is a non-zero scalar multiple of a linear or conjugate-linear isometry.*

Proof   Let L be an injective semilinear transformation of H sending orthogonal vectors to orthogonal vectors. If $x, y \in H$ are orthogonal unit vectors, then $x + y, x - y$ are orthogonal. Since $L(x), L(y)$ and $L(x) + L(y), L(x) - L(y)$ are pairs of orthogonal vectors, we have

$$\|L(x)\| = \|L(y)\|.$$

If unit vectors $x, y \in H$ are non-orthogonal, then there is a unit vector z orthogonal to both x, y (since $\dim H \geq 3$) and we get

$$\|L(x)\| = \|L(z)\| = \|L(y)\|.$$

So, the function $x \to \|L(x)\|$ is constant on the set of unit vectors, which means that L is bounded. Then L is linear or conjugate-linear.

If $\{e_i\}_{i \in I}$ is an orthonormal basis of H, then there is an orthonormal basis $\{e'_i\}_{i \in I}$ of $L(H)$ and non-zero scalars $\{a_i\}_{i \in I}$ such that $L(e_i) = a_i e'_i$. It was established above that $|a_i| = |a_j|$ for any pair $i, j \in I$. Then there is a positive real number b such that for every $i \in I$ we have $a_i = bb_i$ and $|b_i| = 1$. The linear or conjugate-linear operator transferring every e_i to $b_i e'_i$ is an isometry. We have $L = bL'$, where L' is one of these operators.   □

Remark 4.3   Proposition 4.2 holds for the case when $\dim H = 2$ (an exercise for the reader).

4.2 Orthogonality Preserving Transformations

We say that a transformation f of a subset $X \subset \mathcal{L}(H)$ is *orthogonality preserving in both directions* if

$$X \perp Y \iff f(X) \perp f(Y)$$

for any $X, Y \in X$. Every unitary or anti-unitary operator on H induces an automorphism of the lattice $\mathcal{L}(H)$ which preserves the orthogonality relation in both directions.

Theorem 4.4   *Every bijective transformation of $\mathcal{L}(H)$ preserving the orthogonality relation in both directions is a lattice automorphism induced by a unitary or anti-unitary operator on H (such a unitary or anti-unitary operator is unique up to a scalar multiple of modulo one).*

Proof Let f be a bijective transformation of $\mathcal{L}(H)$ preserving the orthogonality relation in both directions. For every $X \in \mathcal{L}(H)$ we denote by ort(X) the set of all elements from $\mathcal{L}(H)$ orthogonal to X. Then

$$f(\text{ort}(X)) = \text{ort}(f(X)).$$

If $X, Y \in \mathcal{L}(H)$, then

$$X \subset Y \Leftrightarrow \text{ort}(Y) \subset \text{ort}(X) \Leftrightarrow \text{ort}(f(Y)) \subset \text{ort}(f(X)) \Leftrightarrow f(X) \subset f(Y).$$

Therefore, f is an automorphism of the lattice $\mathcal{L}(H)$. Every automorphism of this lattice is induced by a semilinear automorphism of H (see Theorem 2.3 and Theorem 3.8 for the finite-dimensional and infinite-dimensional case, respectively). Such a semilinear automorphism sends orthogonal vectors to orthogonal vectors and Proposition 4.2 implies that it is a scalar multiple of a unitary or anti-unitary operator. □

Remark 4.5 Suppose that dim $H = 2$. We take any 1-dimensional subspace $P \subset H$ and consider the transformation of $\mathcal{L}(H)$ which transposes $P, P^\perp$ and leaves fixed all remaining elements of $\mathcal{L}(H)$. This is an automorphism of $\mathcal{L}(H)$ preserving the orthogonality relation in both directions.

Using the same arguments and Theorem 3.17, we prove the following.

Theorem 4.6 (Šemrl [59]) *If H is infinite-dimensional, then every bijective transformation of $\mathcal{G}_\infty(H)$ preserving the orthogonality relation in both directions can be uniquely extended to a lattice automorphism of $\mathcal{L}(H)$ induced by a unitary or anti-unitary operator on H.*

Proof Let f be a bijective transformation of $\mathcal{G}_\infty(H)$ preserving the orthogonality relation in both directions. As in the proof of Theorem 4.4, we establish that f is an automorphism of the partially ordered set $(\mathcal{G}_\infty(H), \subset)$. By Theorem 3.17, f can be uniquely extended to an automorphism of $\mathcal{L}(H)$, i.e. f is induced by an invertible bounded linear or conjugate-linear operator A on H (unique up to a scalar multiple). The operator A sends orthogonal vectors to orthogonal vectors and Proposition 4.2 gives the claim. □

Remark 4.7 The original proof of this statement (see [59]) is not related to Theorem 3.17.

Proposition 4.8 (Uhlhorn [62]) *Every bijective transformation of $\mathcal{G}_1(H)$ preserving the orthogonality relation in both directions can be uniquely extended to a lattice automorphism of $\mathcal{L}(H)$ induced by a unitary or anti-unitary operator on H.*

Proof Let f be a bijective transformation of $\mathcal{G}_1(H)$ preserving the orthogonality relation in both directions. Show that f is an automorphism of the projective space Π_H.

Let S be a 2-dimensional subspace of H. We take any orthogonal basis $\{e_i\}_{i \in I}$ for $S^\perp$ and denote by P_i the 1-dimensional subspace containing e_i. All $f(P_i)$ are mutually orthogonal and we write S' for the maximal closed subspace orthogonal to them. Since f is orthogonality preserving in both directions, $P \in \mathcal{G}_1(H)$ is contained in S if and only if $f(P)$ is contained in S'. The subspace S' is 2-dimensional (this follows from the fact that f^{-1} maps any collection of mutually orthogonal 1-dimensional subspaces of S' to a collection of mutually orthogonal 1-dimensional subspaces of S). So, f transfers lines to lines. Similarly, we show that f^{-1} sends lines to lines.

Therefore, f is induced by a semilinear automorphism of H. This semilinear automorphism sends orthogonal vectors to orthogonal vectors, i.e. it is a scalar multiple of a unitary or anti-unitary operator. □

Remark 4.9 Suppose that $\dim H = n$ is finite and f is an injective transformation of $\mathcal{G}_1(H)$ such that $f(P), f(Q)$ are orthogonal for any orthogonal $P, Q \in \mathcal{G}_1(H)$. As in the proof of Proposition 4.8, we take a 2-dimensional subspace S, an orthonormal basis of $S^\perp$ and the 1-dimensional subspaces $P_1, \dots, P_{n-2}$ containing the vectors from this basis. If a 1-dimensional subspace P is contained in S, then it is orthogonal to all P_i and $f(P)$ is orthogonal to all $f(P_i)$. The latter means that $f(P)$ is contained in the 2-dimensional subspace orthogonal to all $f(P_i)$. Therefore, f sends every line of Π_H to a subset of a line. The image of f is not contained in a line (since it contains three mutually orthogonal elements). By Corollary 2.7, f is induced by a semilinear injection of H to itself. This semilinear injection sends orthogonal vectors to orthogonal vectors. Since H is finite-dimensional, Proposition 4.2 guarantees that it is a scalar multiple of a unitary or anti-unitary operator on H.

We use Chow's theorem (Theorem 2.15) to extend the latter result on other Grassmannians $\mathcal{G}_k(H)$.

Theorem 4.10 (Györy [27] and Šemrl [59]) *If $\dim H > 2k$, then every bijective transformation of $\mathcal{G}_k(H)$ preserving the orthogonality relation in both directions can be uniquely extended to a lattice automorphism of $\mathcal{L}(H)$ induced by a unitary or anti-unitary operator on H.*

Proof Let f be a bijective transformation of $\mathcal{G}_k(H)$ preserving the orthogonality relation in both directions and $\dim H > 2k$. For $k = 1$ the statement was proved above (Proposition 4.8) and we suppose that $k > 1$.

For every subspace $U \subset H$ we denote by $\langle U]_k$ the set of all k-dimensional subspaces contained in U. If $U \in \mathcal{G}^k(H)$, then $\langle U \rangle_k$ consists of all k-dimensional subspaces orthogonal to $U^\perp \in \mathcal{G}_k(H)$. Hence $f(\langle U \rangle_k)$ is formed by all k-dimensional subspaces orthogonal to $f(U^\perp)$, which means that

$$f(\langle U \rangle_k) = \langle g(U) \rangle_k, \quad \text{where} \quad g(U) = f(U^\perp)^\perp.$$

Note that g is a bijective transformation of $\mathcal{G}^k(H)$. The condition $\dim H > 2k$ guarantees that $\mathcal{G}^k(H) \neq \mathcal{G}_k(H)$.

Let $\mathcal{G}'$ be the set of all closed subspaces of H whose codimension is a finite number not less than k and whose dimension is greater than k. If H is infinite-dimensional, then $\mathcal{G}'$ is formed by all closed subspaces of finite codimension $\geq k$. In the case when $\dim H = n$ is finite, it consists of all subspaces $X \subset H$ satisfying

$$k < \dim X \leq n - k$$

(this set is non-empty, since $n > 2k$). We want to extend the bijective transformation g on $\mathcal{G}'$. If U belongs to $\mathcal{G}'$, then $\langle U \rangle_k$ is non-empty and we claim that there exists $g(U) \in \mathcal{G}'$ such that

$$f(\langle U \rangle_k) = \langle g(U) \rangle_k. \tag{4.2}$$

Indeed, U can be presented as the intersection of some $U_1, \ldots, U_i \in \mathcal{G}^k(H)$ and

$$g(U) = g(U_1) \cap \cdots \cap g(U_i)$$

is as required. We get a bijective transformation g of $\mathcal{G}'$ satisfying (4.2) for every $U \in \mathcal{G}'$. Applying the same arguments to f^{-1}, we establish that g is an automorphism of the partially ordered set $(\mathcal{G}', \subset)$; in particular, g preserves the codimensions of subspaces.

Let X and Y be elements of $\mathcal{G}_k(H)$ such that $(X + Y)^\perp$ belongs to $\mathcal{G}'$ (since $\dim H > 2k$, any two adjacent elements of $\mathcal{G}_k(H)$ satisfy this condition). Then $\langle (X+Y)^\perp \rangle_k$ consists of all k-dimensional subspaces orthogonal to both X, Y and

$$f(\langle (X + Y)^\perp \rangle_k) = \langle g((X + Y)^\perp) \rangle_k$$

is formed by all k-dimensional subspaces orthogonal to both $f(X), f(Y)$. This implies that

$$g((X + Y)^\perp) = (f(X) + f(Y))^\perp.$$

Therefore,

$$(X + Y)^\perp \quad \text{and} \quad (f(X) + f(Y))^\perp$$

are of the same finite codimension (since g preserves the codimensions of subspaces). Observe that X, Y are adjacent if and only if the codimension of

$(X + Y)^\perp$ is equal to $k + 1$. Therefore, X, Y are adjacent if and only if $f(X), f(Y)$ are adjacent; i.e. f is an automorphism of the Grassmann graph $\Gamma_k(H)$. It follows from Theorem 2.15 that f is induced by a semilinear automorphism of H. This semilinear automorphism transfers orthogonal vectors to orthogonal vectors and, by Proposition 4.2, it is a scalar multiple of a unitary or anti-unitary operator. $\square$

Remark 4.11   The original proofs from [27, 59] are not related to Chow's theorem. Another proof based on Chow's theorem can be found in [24].

Remark 4.12   We do not consider the case when $\dim H < 2k$, since there are no orthogonal pairs of k-dimensional subspaces in this case. Suppose that $\dim H = 2k$. Then for every $X \in \mathcal{G}_k(H)$ the orthogonal complement $X^\perp$ is the unique k-dimensional subspace orthogonal to X. Consider the set of all pairs $X, X^\perp$, where $X \in \mathcal{G}_k(H)$. Every bijective transformation of this set can be extended to a bijective transformation of $\mathcal{G}_k(H)$ preserving the orthogonality relation in both directions (such an extension is not unique). Therefore, Theorem 4.10 fails for $\dim H = 2k$.

Every linear or conjugate-linear isometry of H to itself induces a transformation of $\mathcal{G}_k(H)$ which preserves the orthogonality relation in both directions. In the case when H is infinite-dimensional, for every natural k there are non-surjective transformations of $\mathcal{G}_k(H)$ which are not induced by linear or conjugate-linear isometries and preserve the orthogonality relation in both directions.

Example 4.13 (Šemrl [59])   Consider a non-surjective linear isometry A of H to itself and fix $X \in \mathcal{G}_k(H)$. Then A^* is surjective and its kernel is the orthogonal complement of $\mathrm{Im}(A)$. Since A^*A is identity, there exists $X' \in \mathcal{G}_k(H)$ which is not contained in $\mathrm{Im}(A)$ and such that $A^*(X') = X$. Consider the transformation f of $\mathcal{G}_k(H)$ defined as follows:

$$f(X) = X' \text{ and } f(Y) = A(Y) \text{ if } Y \in \mathcal{G}_k(H) \setminus \{X\}.$$

Since the image of f is not formed by all k-dimensional subspaces of a certain closed subspace of H, the transformation f cannot be induced by a semilinear isometry. Show that f is orthogonality preserving in both directions. It is clear that $Y, Z \in \mathcal{G}_k(H) \setminus \{X\}$ are orthogonal if and only if $f(Y) = A(Y)$ and $f(Z) = A(Z)$ are orthogonal. For $Y \in \mathcal{G}_k(H) \setminus \{X\}$ the subspaces $f(X) = X'$ and $f(Y) = A(Y)$ are orthogonal if and only if $A^*(X') = X$ and Y are orthogonal.

Suppose that $\dim H$ is finite and greater than $2k$. In this case, we can show that an arbitrary transformation of $\mathcal{G}_k(H)$ preserving the orthogonality relation in both directions is a bijection which can be uniquely extended to a lattice automorphism of $\mathcal{L}(H)$ induced by a unitary or anti-unitary operator on H (see Theorem 4.29). It is natural to ask if the same holds for transformations which send orthogonal subspaces to orthogonal subspaces.

4.3 Non-bijective Version of Wigner's Theorem

In this section, as an exception, we suppose that H is a complex Hilbert space of dimension not less than two.

The *angle* $\angle(P, Q)$ between 1-dimensional subspaces $P, Q \subset H$ is defined as

$$\arccos(|\langle x, y\rangle|),$$

where $x \in P$ and $y \in Q$ are unit vectors (for other unit vectors $x' \in P$ and $y' \in Q$ we have $|\langle x, y\rangle| = |\langle x', y'\rangle|$, which means that the definition does not depend on the choice of unit vectors of P and Q).

Note that $|\langle x, y\rangle|^2$ (the square of the cosine of this angle) is the unique non-zero eigenvalue for the composition of the projections on P and Q. Indeed, the orthogonal projection of the vector x on Q is $\langle x, y\rangle y$ and the orthogonal projection of the latter vector on P is $\langle x, y\rangle\langle y, x\rangle x = |\langle x, y\rangle|^2 x$.

Every transformation of $\mathcal{G}_1(H)$ induced by a linear or conjugate linear isometry preserves the angles between 1-dimensional subspaces.

Example 4.14 Let $\dim H = 2$. For any 1-dimensional subspaces $P, Q \subset H$ there is a unitary operator on H transferring P to $P^\perp$ and Q to $Q^\perp$ and we get

$$\angle(P, Q) = \angle(P^\perp, Q^\perp).$$

Also, we have

$$\angle(P, Q^\perp) = \frac{\pi}{2} - \angle(P, Q).$$

Indeed, if $e_1 \in Q$, $e_2 \in Q^\perp$ and $x = ae_1 + be_2 \in P$ are unit vectors, then

$$|\langle x, e_1\rangle|^2 + |\langle x, e_2\rangle|^2 = |a|^2 + |b|^2 = 1,$$

i.e. the sum of the squares of the cosines of $\angle(P, Q)$ and $\angle(P, Q^\perp)$ is equal to one.

The *gap* between two closed subspaces $X, Y \subset H$ is defined as

$$g(X, Y) = \|P_X - P_Y\|.$$

The function g satisfies the triangle inequality [32, Chapter IV, Section 2.1], i.e. $\mathcal{L}(H)$ together with g is a metric space. This metric space is complete (by the completeness of the operator norm for Banach spaces). The inequality $g(X, Y) < 1$ implies that X and Y are of the same dimension [32, Chapter IV, Corollary 2.6]. Therefore, every subset consisting of all closed subspaces of the same dimension is open; on the other hand, this subset is closed, since our metric space is the disjoint union of all such subsets. In particular, every Grassmannian $\mathcal{G}_k(H)$ is a complete metric space with respect to the gap metric.

Let $P, Q \in \mathcal{G}_1(H)$ and let θ be the angle between P and Q. For unit vectors $x \in P$ and $y \in Q$ denote by x' and y' the orthogonal projections of x and y on Q and P, respectively. Then

$$\|x - x'\| = \sqrt{1 - \|x'\|^2} = \sin \theta = \sqrt{1 - \|y'\|^2} = \|y - y'\|$$

and $g(P, Q) = \sin \theta$.

The classic Wigner's theorem states that every bijective transformation of $\mathcal{G}_1(H)$ preserving the angles is induced by a unitary or anti-unitary operator. In the case when $\dim H \geq 3$, this is a direct consequence of Proposition 4.8, since two elements of $\mathcal{G}_1(H)$ are orthogonal if and only if the angle between them is $\pi/2$.

Theorem 4.15   *Let f be a transformation of $\mathcal{G}_1(H)$ preserving the angles, i.e.*

$$\angle(f(P), f(Q)) = \angle(P, Q)$$

for all $P, Q \in \mathcal{G}_1(H)$. Then f is induced by a linear or conjugate linear isometry of H to itself (such an isometry is unique up to a scalar multiple of modulo one).

Following [23], we present an elementary proof of this statement. Other proofs of Theorem 4.15 can be found in [4] or [37, Section 2.1]. In the case when $\dim H \geq 3$, the statement easily follows from the generalized Fundamental Theorem of Projective Geometry (Corollary 2.7) and [37, Lemma 2.1.2] (see Section 4.8 for the details).

First, we prove Theorem 4.15 for the separable case. Suppose that H is separable and $\{e_i\}_{i \in I}$ is an orthonormal basis of H. Then the set I is $\mathbb{N}$ or $\{1, \dots, n\}$. Let P_i, P'_i, P''_i be the 1-dimensional subspaces containing the vectors

$$e_i, \quad e_i - e_{i+1}, \quad e_i + \mathrm{i}e_{i+1},$$

respectively (if $I = \{1, \dots, n\}$, then P'_i and P''_i are defined only for $i < n$). Denote by $\mathcal{X}$ the set formed by all P_i, P'_i, P''_i and write $\mathcal{D}$ for the set of all 1-dimensional subspaces non-orthogonal to all P_i.

Lemma 4.16   *If $Q, Q' \in \mathcal{D}$ and $\angle(Q, P) = \angle(Q', P)$ for every $P \in \mathcal{X}$, then $Q = Q'$.*

Proof Let

$$x = \sum_{i \in I} a_i e_i \quad \text{and} \quad x' = \sum_{i \in I} a'_i e_i$$

be unit vectors belonging to Q and Q', respectively. Since $Q, Q' \in \mathcal{D}$, all a_i and a'_i are non-zero. If the equality $\angle(Q, P) = \angle(Q', P)$ holds for every $P \in X$, then

$$|a_i| = |a'_i|, \quad |a_i - a_{i+1}| = |a'_i - a'_{i+1}|, \quad |a_i - \mathrm{i}a_{i+1}| = |a'_i - \mathrm{i}a'_{i+1}|$$

for all $i \in I$. Without loss of generality we can assume that $a_1 = a'_1$.

Suppose that $a_i = a'_i$ for all i not greater than a certain k. Using the equalities

$$|a_{k+1}| = |a'_{k+1}|, \quad |a_k - a_{k+1}| = |a_k - a'_{k+1}|, \quad |a_k - \mathrm{i}a_{k+1}| = |a_k - \mathrm{i}a'_{k+1}|,$$

we establish that $a_{k+1} = a'_{k+1}$ (we leave all details for the reader). So, every a_i coincides with a'_i and we get $Q = Q'$.   □

Remark 4.17 The above statement fails for an arbitrary pair $Q, Q' \in \mathcal{G}_1(H)$. Consider, for example, the 1-dimensional subspaces containing the vectors $e_1 + e_3$ and $e_1 + \mathrm{i}e_3$.

Proof of Theorem 4.15 in the separable case Since f preserves the angles between 1-dimensional subspaces, it is orthogonality preserving in both directions. This means that $\{f(P_i)\}_{i \in I}$ are mutually orthogonal and we denote by W the closed subspace of H spanned by them. There is a linear isometry $A : W \to H$ transferring every $f(P_i)$ to P_i. Then Af is a transformation of $\mathcal{G}_1(H)$ preserving the angles and leaving fixed every P_i. For this reason, we can assume that f leaves fixed each P_i.

Then the equalities

$$\angle(f(P'_i), P_j) = \angle(P'_i, P_j) \quad \text{and} \quad \angle(f(P''_i), P_j) = \angle(P''_i, P_j)$$

show that $f(P'_i)$ and $f(P''_i)$ contain vectors

$$e_i - c_{i+1} e_{i+1} \quad \text{and} \quad e_i - d_{i+1} e_{i+1}$$

(respectively), where $|c_{i+1}| = |d_{i+1}| = 1$. Since

$$\angle(f(P'_i), f(P''_i)) = \angle(P'_i, P''_i),$$

we have

$$\sqrt{2} = |\langle e_i - e_{i+1}, e_i + \mathrm{i}e_{i+1} \rangle| = |\langle e_i - c_{i+1} e_{i+1}, e_i - d_{i+1} e_{i+1} \rangle| = |1 - c_{i+1}\overline{d}_{i+1}|,$$

which implies that $d_{i+1} = \pm \mathrm{i}c_{i+1}$.

If $d_2 = -ic_2$, then we take the unitary operator U satisfying

$$U(e_1) = e_1 \quad \text{and} \quad U(e_i) = c_2 \cdots c_i e_i \quad \text{for } i \geq 2.$$

In the case when $d_2 = ic_2$, we suppose that U is the anti-unitary operator satisfying the same conditions. In each of these cases, $U^* f$ is a transformation of $\mathcal{G}_1(V)$ preserving the angles and leaving fixed all P_i and P_i'. Also, this transformation leaves fixed P_1'' and sends every P_i'', $i \geq 2$ to the 1-dimensional subspace containing $e_i \pm ie_{i+1}$. Therefore, we can assume that f satisfies the following conditions:

- f leaves fixed all P_i, P_i' and P_1'',
- $f(P_i'')$ contains $e_i \pm ie_{i+1}$ if $i \geq 2$.

Suppose that j is the first index such that $f(P_j'')$ contains $e_j - ie_{j+1}$ and show that this assumption gives a contradiction.

Let Q be the 1-dimensional subspace containing a unit vector

$$ae_{j-1} + te_j + be_{j+1},$$

where $t > 0$ and a, b are non-zero. Then $f(Q)$ contains a unit vector

$$a'e_{j-1} + se_j + b'e_{j+1}$$

such that $s > 0$. It is easy to see that $s = t$. The equalities

$$|a| = |a'|, \quad |a - t| = |a' - t|, \quad |a - it| = |a' - it|$$

imply that $a = a'$. Using the equalities

$$|b| = |b'|, \quad |b - t| = |b' - t|, \quad |b - it| = |b' + it|,$$

we establish that $b' = \overline{b}$.

Now, we consider the 1-dimensional subspaces Q and Q' containing the vectors

$$-\frac{1}{2}e_{j-1} + \frac{1}{2}e_j + \frac{1}{\sqrt{2}}e_{j+1} \quad \text{and} \quad \frac{i}{2}e_{j-1} + \frac{1}{2}e_j + \frac{i}{\sqrt{2}}e_{j+1},$$

respectively. Then $f(Q) = Q$ and $f(Q')$ contains the unit vector

$$\frac{i}{2}e_{j-1} + \frac{1}{2}e_j - \frac{i}{\sqrt{2}}e_{j+1}.$$

A direct calculation shows that the cosines of $\angle(Q, Q')$ and $\angle(f(Q), f(Q'))$ are equal to $\sqrt{2}/4$ and $\sqrt{10}/4$ (respectively), which gives a contradiction.

So, f leaves fixed every element of X and, by Lemma 4.16, it also leaves fixed all elements of $\mathcal{D}$. Consider $\mathcal{G}_1(H)$ as the metric space, where the distance between $P, Q \in \mathcal{G}_1(H)$ is equal to the gap $g(P, Q)$, i.e the sine of the angle

$\angle(P, Q)$. The set $\mathcal{D}$ is everywhere dense in this metric space (as the countable intersection of open everywhere dense subsets of a complete metric space). Since f is a continuous transformation of this metric space (it preserves the angles, and consequently the gap metric), f is identity. $\qquad\square$

Proof of Theorem 4.15 in the non-separable case Consider the case when H is non-separable. Let

$$\{e_1, e_2\} \cup \{e_{t,i} : t \in T \text{ and } i \in \mathbb{N}, i \geq 3\}$$

be an orthonormal basis of H. In what follows, we will suppose that $e_{t,1} = e_1$ and $e_{t,2} = e_2$ for every $t \in T$. As in the separable case, we can assume that f leaves fixed all 1-dimensional subspaces containing the basis vectors.

For every $t \in T$ we denote by H_t the closed subspace spanned by $\{e_{t,i}\}_{i\in\mathbb{N}}$. Then for any distinct $t, t' \in T$ the intersection of H_t and $H_{t'}$ is the 2-dimensional subspace S containing the vectors e_1 and e_2. Since f is orthogonality preserving and leaves fixed all 1-dimensional subspaces containing the basis vectors,

$$f(\mathcal{G}_1(S)) \subset \mathcal{G}_1(S) \text{ and } f(\mathcal{G}_1(H_t)) \subset \mathcal{G}_1(H_t)$$

for every $t \in T$. There is a unitary or anti-unitary operator U_t on H_t such that $U_t f$ leaves fixed every element of $\mathcal{G}_1(H_t)$ and we can assume that $U_t(e_1) = e_1$. All $U_t|_S$ define the same transformation of $\mathcal{G}_1(S)$, which means that $U_t|_S$ and $U_{t'}|_S$ are proportional for any $t, t' \in T$ and, by the assumption that $U_t(e_1) = e_1$, we have

$$U_t|_S = U_{t'}|_S.$$

Therefore, there is a unitary or anti-unitary operator U on H such that Uf leaves fixed all elements of $\mathcal{G}_1(H_t)$ for every $t \in T$. In other words, we can assume that the restriction of f to each $\mathcal{G}_1(H_t)$ is identity.

Consider $\mathcal{G}_1(H)$ as the metric space with respect to the gap metric. Since f is continuous and the set of all 1-dimensional subspaces of H non-orthogonal to e_1 is everywhere dense, it is sufficient to show that f leaves fixed every such 1-dimensional subspace.

Let P be a 1-dimensional subspace of H non-orthogonal to e_1 and let $x \in P$ and $y \in f(P)$ be unit vectors. For every $t \in T$ we denote by x_t and y_t the orthogonal projections of these vectors on H_t. We have

$$\angle(P, Q) = \angle(f(P), Q)$$

for all $Q \in \mathcal{G}_1(H_t)$, i.e.

$$|\langle x_t, x \rangle| = |\langle y_t, x \rangle|$$

for every unit vector $x \in H_t$. This is possible only in the case when $y_t = a_t x_t$

and a_t is a scalar of modulo one. All vectors x_t have the same non-zero e_1-coordinate and all y_t also have the same non-zero e_1-coordinate. Therefore, $a_t = a_{t'}$ for any $t, t' \in T$ and y is a scalar multiple of x, which implies that $f(P) = P$. □

4.4 The Principal Angles between Subspaces

The principal angles between finite-dimensional subspaces were introduced in [29]. Following [5, Section VII.1] (see also [10]) we give the definition of the principal angles based on the cosine–sine decomposition.

Theorem 4.18 *Let X and Y be k-dimensional subspaces of H. Then there exist orthonormal bases $x_1, \ldots, x_k$ and $y_1, \ldots, y_k$ of X and Y (respectively) such that x_i and y_j are orthogonal if $i \neq j$ and every $\langle x_i, y_i \rangle$ is a non-negative real number.*

Proof We will use the CS decomposition of unitary matrices (see, for example, [5, Theorem VII.1.6]): for every unitary matrix U of order $2k$ there are order-k unitary matrices A_1, A_2, B_1, B_2 such that

$$\begin{pmatrix} A_1 & 0 \\ 0 & A_2 \end{pmatrix} U \begin{pmatrix} B_1 & 0 \\ 0 & B_2 \end{pmatrix} = \begin{pmatrix} C & -S \\ S & C \end{pmatrix},$$

where C and S are diagonal order-k matrices with real diagonal entries

$$1 \geq c_1 \geq \cdots \geq c_k \geq 0 \quad \text{and} \quad 0 \leq s_1 \leq \cdots \leq s_k \leq 1,$$

respectively, and $C^2 + S^2 = I_k$ (I_k is the identity matrix of order k).

First, we consider the case when $X \cap Y = 0$. We take $(n \times k)$ - matrices

$$M = \begin{pmatrix} M_1 \\ M_2 \end{pmatrix} \quad \text{and} \quad N = \begin{pmatrix} N_1 \\ N_2 \end{pmatrix}$$

whose columns correspond to vectors in orthonormal bases of X and Y, respectively; note that M_1, M_2, N_1, N_2 are $(k \times k)$ - matrices. Without loss of generality, we can assume that $M_1 = I_k$ and $M_2 = 0$. We extend N to an order-$2k$ unitary matrix

$$\begin{pmatrix} N_1 & * \\ N_2 & * \end{pmatrix}$$

(this is possible, since the columns of N are mutually orthogonal). It was noted above that there are order-k unitary matrices A_1, A_2, B_1, B_2 such that

$$\begin{pmatrix} A_1 & 0 \\ 0 & A_2 \end{pmatrix} \begin{pmatrix} N_1 & * \\ N_2 & * \end{pmatrix} \begin{pmatrix} B_1 & 0 \\ 0 & B_2 \end{pmatrix} = \begin{pmatrix} C & -S \\ S & C \end{pmatrix}.$$

Then

$$\begin{pmatrix} A_1 N_1 B_1 & * \\ A_2 N_2 B_1 & * \end{pmatrix} = \begin{pmatrix} C & -S \\ S & C \end{pmatrix}$$

and from the first columns of the left and right sides we obtain that

$$A_1 N_1 B_1 = C \quad \text{and} \quad A_2 N_2 B_1 = S.$$

This implies that

$$AMA_1^* = \begin{pmatrix} I_k \\ 0 \end{pmatrix} \quad \text{and} \quad ANB_1 = \begin{pmatrix} C \\ S \end{pmatrix},$$

where

$$A = \begin{pmatrix} A_1 & 0 \\ 0 & A_2 \end{pmatrix}.$$

The left multiplying M and N by the order-$2k$ unitary matrix A corresponds to a unitary automorphism of $X + Y$. The right multiplying M and N by the order-k unitary matrices A_1^* and B_1 (respectively) corresponds to a change of bases in the subspaces X and Y. Therefore, there exist orthonormal bases $x_1, \ldots, x_k$ and $y_1, \ldots, y_k$ for X and Y (respectively) such that x_i and y_j are orthogonal if $i \neq j$ and $\langle x_i, y_i \rangle = c_i$ for each i.

In the case when $X \cap Y \neq 0$, we apply the above arguments to the subspaces $X \cap (X \cap Y)^\perp$ and $Y \cap (X \cap Y)^\perp$. □

As above, we suppose that X and Y are k-dimensional subspaces of H. Let $x_1, \ldots, x_k$ and $y_1, \ldots, y_k$ be orthonormal bases for X and Y (respectively) such that x_i and y_j are orthogonal if $i \neq j$. Permuting indices and replacing basis vectors by scalar multiples of modulo one, we obtain that each $\langle x_i, y_i \rangle = c_i$ is a real number and

$$1 \geq c_1 \geq \cdots \geq c_k \geq 0.$$

Consider the self-adjoint operators $P_X P_Y P_X$ and $P_Y P_X P_Y$. Observe that c_i^2 is the eigenvalue of both the operators with respect to the eigenvectors x_i and y_i, respectively. If $x_1', \ldots, x_k' \in X$ and $y_1', \ldots, y_k' \in Y$ are orthonormal bases satisfying $\langle x_i', y_j' \rangle = 0$ for $i \neq j$, then x_i' and y_i' are eigenvectors of $P_X P_Y P_X$ and $P_Y P_X P_Y$ (respectively) corresponding to the eigenvalue $|\langle x_i', y_i' \rangle|^2$ and there is a permutation σ on the set $\{1, \ldots, k\}$ such that

$$|\langle x_i', y_i' \rangle| = c_{\sigma(i)}$$

for every i. We define the collection of the *principal angles*

$$\angle(X, Y) = \{\theta_1, \ldots, \theta_k\}, \quad 0 \leq \theta_1 \leq \cdots \leq \theta_k \leq \pi/2$$

between X and Y as

$$\theta_i = \arccos(c_i) = \arccos(|\langle x_i, y_i \rangle|)$$

for every $i \in \{1, \dots, k\}$, i.e. θ_i is the angle between the 1-dimensional subspaces containing x_i and y_i. The definition does not depend on the choice of orthonormal bases $x_1, \dots, x_k \in X$ and $y_1, \dots, y_k \in Y$ satisfying $\langle x_i, y_j \rangle = 0$ for $i \neq j$.

The subspaces X and Y are orthogonal if and only if $\theta_i = \pi/2$ for every i. In the case when

$$\dim(X \cap Y) = m > 0,$$

every $x \in X \cap Y$ is an eigenvector of both $P_X P_Y P_X$ and $P_Y P_X P_Y$ corresponding to the eigenvalue 1, i.e. $x_i = y_i$ and $\theta_i = 0$ for every $i \leq m$. The subspaces X and Y are compatible if and only if each θ_i is equal to 0 or $\pi/2$. Recall that X, Y are *adjacent* if their intersection is $(k-1)$-dimensional. This is equivalent to the fact that only one of the principal angles between X and Y is non-zero. In the case when this angle is $\pi/2$, we say that X and Y are *ortho-adjacent*. In other words, two elements of $\mathcal{G}_k(H)$ are ortho-adjacent if they are adjacent and compatible.

Remark 4.19 If $x = \sum_{i=1}^{k} a_i x_i \in X$ is a unit vector and x' is its orthogonal projection on Y, then

$$\|x'\|^2 = \sum_{i=1}^{k} |a_i|^2 \cos^2 \theta_i \geq \cos^2 \theta_k$$

and $\|x'\| = \cos \theta_k$ if $x = x_k$. This means that for unit vectors $x \in X$ the function

$$\|x - x'\| = \sqrt{1 - \|x'\|^2}$$

takes the maximum value at $x = x_k$. Similarly, if $y \in Y$ is a unit vector and y' is its orthogonal projection on X, then $\|y - y'\|$ takes the maximum value at $y = y_k$. Therefore, $g(X, Y) = \sin \theta_k$.

Remark 4.20 Now, we describe some properties of the principal angles which are not exploited in what follows (for this reason, we leave all the details as exercises for the reader). Denote by X_i and Y_i the eigenspaces of $P_X P_Y P_X$ and $P_Y P_X P_Y$ (respectively) corresponding to the eigenvalue $\cos^2 \theta_i$ (in the case when $\theta_i = \pi/2$, we do not include in X_i and Y_i vectors which do not belong to X and Y, respectively). If $\theta_i = \theta_j$, then $X_i = X_j$ and $Y_i = Y_j$. Let $x_1, \dots, x_k \in X$ and $y_1, \dots, y_k \in Y$ be orthonormal bases satisfying $|\langle x_i, y_j \rangle| = \theta_i$ for every i and $\langle x_i, y_j \rangle = 0$ for $i \neq j$. Then X_i is spanned by all x_j such that $\theta_i = \theta_j$ and the same holds for Y_i, i.e. X_i and Y_i are of the same dimension. Let P and Q be

1-dimensional subspaces contained in the subspaces spanned by $x_i, \ldots, x_n$ and $y_i, \ldots, y_n$, respectively. Using the Cauchy–Schwarz inequality we can prove the following:

(A) $\angle(P, Q) \geq \theta_i$ and the equality $\angle(P, Q) = \theta_i$ implies that $P \subset X_i$ and $Q \subset Y_i$.

Suppose that X_i and Y_i are spanned by $x_i, \ldots, x_t$ and $y_i, \ldots, y_t$ (respectively). For unit vectors

$$x = \sum_{j=i}^{t} a_j x_j \in X_i \text{ and } y = \sum_{j=i}^{t} b_j y_j \in Y_i$$

the equality $|\langle x, y \rangle| = \cos \theta_i$ means that

$$\left| \sum_{j=i}^{t} a_j \overline{b}_j \right| = 1.$$

Replacing x_j and y_j by scalar multiples of modulo one, we reduce the general case to the case when all a_j, b_j are non-negative real numbers. Then

$$\sum_{j=i}^{t} a_j b_j = \sum_{j=i}^{t} a_j^2 = \sum_{j=i}^{t} b_j^2 = 1,$$

which implies that $a_j = b_j$ for all j. Therefore, we get the following:

(B) If $\angle(P, Q) = \theta_i$, then $P_Y(P) = Q$ and $P_X(Q) = P$.

These facts show that the collection of the principal angles between X and Y can be constructed recursively. By the statement (A), θ_1 is the minimum value of $\arccos(|\langle x, y \rangle|)$ for unit vectors $x \in X$ and $y \in Y$. Let $x_1' \in X$ and $y_1' \in Y$ be any two unit vectors realizing this minimum. Now, we define the angles $\theta_2', \ldots, \theta_k'$ and the corresponding unit vectors $x_2', \ldots, x_k' \in X$ and $y_2', \ldots, y_k' \in Y$. The ith angle θ_i' is the minimum value of $\arccos(|\langle x, y \rangle|)$ for unit vectors $x \in X$ and $y \in Y$ orthogonal to

$$x_1', \ldots, x_{i-1}' \text{ and } y_1', \ldots, y_{i-1}'$$

(respectively) and we take unit vectors $x_i' \in X$, $y_i' \in Y$ satisfying the latter conditions and realizing this minimum. Using (A) and (B), we establish that $\theta_i' = \theta_i$ for all $i \geq 2$; moreover, $|\langle x_i', y_i' \rangle| = \cos \theta_i$ for every i and $\langle x_i', y_j' \rangle = 0$ if $i \neq j$.

For k-dimensional subspaces $X, Y \subset H$ we define

$$\text{tr}(X, Y) = \cos^2 \theta_1 + \cdots + \cos^2 \theta_k,$$

where $\angle(X, Y) = \{\theta_1, \ldots, \theta_k\}$. This is the trace of the operators $P_X P_Y P_X$ and $P_Y P_X P_Y$.

Proposition 4.21   *Suppose that* $\dim H = n$ *is finite. If* X *and* Y *are* k-*dimensional subspaces of* H, *then*

$$\mathrm{tr}(X^{\perp}, Y^{\perp}) - \mathrm{tr}(X, Y) = n - 2k.$$

In the case when $n = 2k$, *the following assertions are fulfilled:*

- $\angle(X^{\perp}, Y^{\perp}) = \angle(X, Y)$,
- $\angle(X, Y^{\perp}) = \left\{\frac{1}{2}\pi - \theta_k, \ldots, \frac{1}{2}\pi - \theta_1\right\}$ *if* $\angle(X, Y) = \{\theta_1, \ldots, \theta_k\}$.

Proof   Let $x_1, \ldots, x_k \in X$ and $y_1, \ldots, y_k \in Y$ be orthonormal bases such that $|\langle x_i, y_j \rangle| = \theta_i$ for every i and $\langle x_i, y_j \rangle = 0$ for $i \neq j$. If $\dim(X \cap Y) = m$, then $x_i = y_i$ for every $i \leq m$ and

$$\angle(X, Y) = \{\underbrace{0, \ldots, 0}_{m}, \theta_{m+1}, \ldots, \theta_k\},$$

where $\theta_i > 0$ for all $i > m$. For each $i > m$ we denote by S_i the 2-dimensional subspace spanned by x_i and y_i. The subspaces

$$X \cap Y, \ (X + Y)^{\perp}, \ S_{m+1}, \ldots, S_k$$

are mutually orthogonal and their sum coincides with H. We take unit vectors $x_i', y_i' \in S_i$ orthogonal to x_i and y_i, respectively. Let B be an orthonormal basis of $X^{\perp} \cap Y^{\perp} = (X + Y)^{\perp}$. Then

$$B \cup \{x_{m+1}', \ldots, x_k'\} \quad \text{and} \quad B \cup \{y_{m+1}', \ldots, y_k'\}$$

are orthonormal bases of $X^{\perp}$ and $Y^{\perp}$, respectively. Each x_i' is orthogonal to all vectors from B and to y_j' if $i \neq j$. By Example 4.14,

$$|\langle x_i', y_i' \rangle| = |\langle x_i, y_i \rangle| = \theta_i$$

for every $i > m$. Since

$$\dim(X^{\perp} \cap Y^{\perp}) = n - 2k + m,$$

we have

$$\angle(X^{\perp}, Y^{\perp}) = \{\underbrace{0, \ldots, 0}_{n-2k+m}, \theta_{m+1}, \ldots, \theta_k\},$$

which implies the first equality.

Suppose that $n = 2k$. Then the dimension of $X^{\perp} \cap Y^{\perp}$ is m and we obtain the second equality. The third equality is a simple consequence of the fact that

$$|\langle x_i, y_i \rangle|^2 + |\langle x_i, y_i' \rangle|^2 = 1$$

for every $i > m$ (Example 4.14). □

Remark 4.22 There is an interesting class of metrics on the Grassmannian $\mathcal{G}_k(H)$ related to the principal angles [53]. A metric belongs to this class if it is defined as

$$d(X, Y) = n(\theta_1, \dots, \theta_k),$$

where $X, Y \in \mathcal{G}_k(H)$, $\angle(X, Y) = \{\theta_1, \dots, \theta_k\}$ and $n : \mathbb{R}^k \to \mathbb{R}$ is a norm function. This holds, for example, for the gap metric (Remark 4.19). Another metric of such type will be considered in Section 5.7.

Remark 4.23 The principal angles can be defined between any two finite-dimensional subspaces of H; the number of angles is equal to the smallest of the dimensions. The angles between closed infinite-dimensional subspaces are described in [33].

4.5 Transformations of Grassmannians Preserving the Principal Angles

Every transformation of $\mathcal{G}_k(H)$ induced by a linear or conjugate-linear isometry preserves the principal angles between subspaces. If $\dim H = 2k$, then Proposition 4.21 shows that the same holds for the restriction of the orthocomplementation to $\mathcal{G}_k(H)$.

Theorem 4.24 (Molnár [36, 38]) *Let f be a transformation of $\mathcal{G}_k(H)$ preserving the principal angles, i.e.*

$$\angle(f(X), f(Y)) = \angle(X, Y)$$

for all $X, Y \in \mathcal{G}_k(H)$. Then one of the following possibilities is realized:

- *f is induced by a linear or conjugate-linear isometry of H to itself (such an isometry is unique up to a scalar multiple of modulo one),*
- *$\dim H = 2k$ and f is a bijection which can be uniquely extended to an automorphism or anti-automorphism of the lattice $\mathcal{L}(H)$ induced by a unitary or anti-unitary operator on H, i.e. there is a unitary or anti-unitary operator U (unique up to a scalar multiple of modulo one) such that $f(X) = U(X)$ for all $X \in \mathcal{G}_k(H)$ or $f(X) = U(X)^\perp$ for all $X \in \mathcal{G}_k(H)$.*

Recall that for any k-dimensional subspaces $X, Y \subset H$

$$\mathrm{tr}(X, Y) = \cos^2 \theta_1 + \cdots + \cos^2 \theta_k,$$

where $\angle(X, Y) = \{\theta_1, \dots, \theta_k\}$. Theorem 4.24 can be generalized as follows.

Theorem 4.25 (Gehér [25])   *Let f be a transformation of $\mathcal{G}_k(H)$ satisfying the following condition:*

$$\mathrm{tr}(f(X), f(Y)) = \mathrm{tr}(X, Y)$$

for all $X, Y \in \mathcal{G}_k(H)$. Then f is induced by a linear or conjugate-linear isometry of H to itself or $\dim H = 2k$ and f is a bijection which can be uniquely extended to an automorphism or anti-automorphism of the lattice $\mathcal{L}(H)$ induced by a unitary or anti-unitary operator on H.

The proof of Theorem 4.25 is presented in Section 4.8.

In the case when $k = 1$, this statement coincides with Theorem 4.15. Suppose that $\dim H = n$ is finite and consider the transformation g of $\mathcal{G}_{n-k}(H)$ defined as

$$g(X) = f(X^{\perp})^{\perp}$$

for every $X \in \mathcal{G}_{n-k}(H)$, where f is from Theorem 4.25. By Proposition 4.21 and the assumption of Theorem 4.25, we have

$$\mathrm{tr}(g(X), g(Y)) = \mathrm{tr}(X, Y)$$

for all $X, Y \in \mathcal{G}_{n-k}(H)$. If g can be extended to an automorphism of $\mathcal{L}(H)$ induced by a unitary or anti-unitary operator, then the restriction of this automorphism to $\mathcal{G}_k(H)$ coincides with f. Therefore, it is sufficient to prove Theorem 4.25 only in the case when $\dim H \geq 2k > 2$.

For the case when $\dim H > 2k$ and f is bijective, Theorem 4.25 (and, consequently, Theorem 4.24) follows from Theorem 4.10.

We present another generalization of Theorem 4.24 which says that it is sufficient to require that a transformation of $\mathcal{G}_k(H)$ preserves only some types of the principal angles to assert that this transformation is induced by a linear or conjugate-linear isometry. It was noted above that the general case is reducible to the case when $\dim H \geq 2k > 2$.

Theorem 4.26 (Pankov [48])   *Suppose that $\dim H > 2k > 2$. Let f be an orthogonality preserving transformation of $\mathcal{G}_k(H)$ which satisfies one of the following additional conditions:*

(A) *f is adjacency preserving;*

(OA) *f is an ortho-adjacency preserving injection.*

Then f is induced by a linear or conjugate-linear isometry of H to itself.

The proof of Theorem 4.26 is given in Section 4.6.

Remark 4.27 Consider the case when $k = 1$. Since any two distinct elements of $\mathcal{G}_1(H)$ are adjacent, the condition (A) is equivalent to the fact that f is injective. By Remark 4.9, the conclusion of Theorem 4.26 holds true if $\dim H$ is finite and not less than three. In the case when H is infinite-dimensional, the conclusion fails by Example 4.13.

For the case when $\dim H = 2k$, we can prove only the following weak version of Theorem 4.26.

Proposition 4.28 (Pankov [48]) *Suppose that* $\dim H = 2k > 2$. *Let f be an orthogonality preserving transformation of $\mathcal{G}_k(H)$ which preserves the adjacency relation in both directions. Then f is a bijection which can be uniquely extended to an automorphism or anti-automorphism of the lattice $\mathcal{L}(H)$ induced by a unitary or anti-unitary operator on H.*

The proof of Proposition 4.28 is also given in Section 4.6.

If H is infinite-dimensional, then for every natural k there are non-surjective transformations of $\mathcal{G}_k(H)$ which are not induced by linear or conjugate-linear isometries and preserve the orthogonality relation in both directions (Example 4.13). In the case when $\dim H$ is finite and greater than $2k$, such transformations do not exist. We use Theorem 4.26 to prove the following.

Theorem 4.29 (Pankov [48]) *If the dimension of H is finite and greater than $2k$, then every transformation of $\mathcal{G}_k(H)$ preserving the orthogonality relation in both directions is a bijection which can be uniquely extended to an automorphism of the lattice $\mathcal{L}(H)$ induced by a unitary or anti-unitary operator on H.*

The proof of Theorem 4.29 is given in Section 4.7.

4.6 Proofs of Theorem 4.26 and Proposition 4.28

Let $\dim H \geq 2k > 2$. We will use some properties of the Grassmann graph $\Gamma_k(H)$ formed by k-dimensional subspaces of H.

Recall that there are precisely two types of maximal cliques in the graph $\Gamma_k(H)$ called stars and tops. If $X \in \mathcal{G}_{k-1}(H)$, then the star $\mathcal{S}(X)$ consists of all elements of $\mathcal{G}_k(H)$ containing X. For $Y \in \mathcal{G}_{k+1}(H)$ the corresponding top is $\mathcal{T}(Y) = \mathcal{G}_k(Y)$.

Also, we recall that the orthogonal apartment of $\mathcal{G}_k(H)$ associated to an orthonormal basis B of H consists of all k-dimensional subspaces spanned by subsets of B. This is a compatible subset, i.e. any two distinct elements from this subset are compatible. By Proposition 1.15, every compatible subset of $\mathcal{G}_k(H)$ is contained in an orthogonal apartment. A maximal compatible subset in a star $\mathcal{S}(X)$ or a top $\mathcal{G}_k(Y)$ is the intersection of this star or top with an orthogonal apartment such that X or Y is spanned by a subset of the corresponding orthonormal basis. The following statement is obvious.

Lemma 4.30 *Every maximal compatible subset of a top contains precisely $k + 1$ elements. Every maximal compatible subset of a star contains precisely $n - k + 1$ elements if $\dim H = n$ is finite, and this set is infinite if H is infinite-dimensional.*

Let X and Y be orthogonal elements of $\mathcal{G}_k(H)$. The distance between them in the graph $\Gamma_k(H)$ is equal to k. If

$$X = X_0, X_1, \ldots, X_k = Y$$

is a geodesic in $\Gamma_k(H)$, then

$$\dim(X \cap X_i) = k - i \quad \text{and} \quad \dim(Y \cap X_i) = i$$

for every $i \in \{1, \ldots, k\}$. This means that X_i is the orthogonal sum of $X \cap X_i$ and $Y \cap X_i$. Therefore, X_i is compatible to both X and Y. If $0 < i < j < k$, then Lemma 2.12 states that $X \cap X_j$ is contained in $X \cap X_i$, and $Y \cap X_i$ is contained in $Y \cap X_j$, which implies that X_i and X_j are compatible (by Lemma 1.14).

Now, let us consider compatible $X, Y \in \mathcal{G}_k(H)$. We take $Z \in \mathcal{G}_k(H)$ intersecting Y precisely in $(X \cap Y)^{\perp} \cap Y$ and orthogonal to X. Then X, Y, Z are mutually compatible and there is an orthogonal apartment of $\mathcal{G}_k(H)$ containing them. This apartment contains a geodesic joining X with Z and passing through Y.

So, we obtain the following characterization of a compatibility relation in terms of orthogonality and adjacency.

Lemma 4.31 *Every geodesic in $\Gamma_k(H)$ joining orthogonal elements consists of mutually compatible elements. Any two compatible $X, Y \in \mathcal{G}_k(H)$ are contained in a certain geodesic of $\Gamma_k(H)$ connecting X with an element orthogonal to X.*

The case (A) *of Theorem 4.26.* Suppose that $\dim H > 2k$. Let f be an orthogonality preserving transformation of $\mathcal{G}_k(H)$ satisfying the condition (A) from Theorem 4.26, i.e. f is adjacency preserving.

Lemma 4.32 *The transformation f is ortho-adjacency preserving.*

Proof Suppose that $X, Y \in \mathcal{G}_k(H)$ are ortho-adjacent. Then $f(X)$ and $f(Y)$ are adjacent and we need to show that they are compatible. By the second statement of Lemma 4.31, Y is contained in a certain geodesic of $\Gamma_k(H)$ connecting

X with an element $Z \in \mathcal{G}_k(H)$ orthogonal to X. Since f is adjacency preserving, it transfers this geodesic to a path in $\Gamma_k(H)$. The elements X and Z are orthogonal and the same holds for $f(X)$ and $f(Z)$. This implies that the image of our geodesic is a geodesic in $\Gamma_k(H)$ connecting $f(X)$ with $f(Z)$ and containing $f(Y)$. Then $f(X)$ and $f(Y)$ are compatible by the first statement of Lemma 4.31. $\square$

Lemma 4.33 *For every star $\mathcal{S} \subset \mathcal{G}_k(H)$ there is a unique star containing $f(\mathcal{S})$.*

Proof It is clear that $f(\mathcal{S})$ is a clique in $\Gamma_k(H)$ (not necessarily maximal), i.e. $f(\mathcal{S})$ is contained in a certain maximal clique (a star or a top). If X is a maximal compatible subset of $\mathcal{S}$, then $f(X)$ is a compatible subset in a star or a top by Lemma 4.32. Since $\dim H > 2k$, Lemma 4.30 shows that $f(X)$ cannot be contained in a top. Therefore, $f(\mathcal{S})$ is a subset in a star. The intersection of two distinct stars contains at most one element. This means that there is a unique star containing $f(\mathcal{S})$. $\square$

Lemma 4.33 shows that f induces a transformation f_{k-1} of $\mathcal{G}_{k-1}(H)$ satisfying

$$f(\mathcal{S}(X)) \subset \mathcal{S}(f_{k-1}(X))$$

for every $X \in \mathcal{G}_{k-1}(H)$. Then

$$f_{k-1}(\mathcal{G}_{k-1}(Y)) \subset \mathcal{G}_{k-1}(f(Y))$$

for every $Y \in \mathcal{G}_k(H)$.

Lemma 4.34 *The transformation f_{k-1} is orthogonality preserving.*

Proof If X and Y are orthogonal elements of $\mathcal{G}_{k-1}(H)$, then there are orthogonal $X', Y' \in \mathcal{G}_k(H)$ such that $X \subset X'$ and $Y \subset Y'$. We have

$$f_{k-1}(X) \subset f(X'), \quad f_{k-1}(Y) \subset f(Y') \quad \text{and} \quad f(X') \perp f(Y'),$$

which implies that $f_{k-1}(X)$ and $f_{k-1}(Y)$ are orthogonal. $\square$

Lemma 4.35 *The following assertions are fulfilled:*

(1) *If $X, Y \in \mathcal{G}_{k-1}(H)$ are adjacent, then $f_{k-1}(X)$ and $f_{k-1}(Y)$ are adjacent or coincident.*
(2) *f_{k-1} is ortho-adjacency preserving.*

Proof The statements are trivial for $k = 2$ and we suppose that $k > 2$.

If $X, Y \in \mathcal{G}_{k-1}(H)$ are adjacent, then the corresponding stars $\mathcal{S}(X)$ and $\mathcal{S}(Y)$

have a non-empty intersection. The transformation f sends these stars to subsets of a star or to subsets of distinct stars with a non-empty intersection. This implies the statement (1).

The statement (1) together with Lemma 4.34 guarantee that f_{k-1} transfers every geodesic of $\Gamma_{k-1}(H)$ connecting orthogonal elements to a geodesic which connects orthogonal elements. Therefore, the statement (2) follows from Lemma 4.31 as in the proof of Lemma 4.32.   □

If $k \geq 3$, then we use Lemma 4.35 and the arguments from the proof of Lemma 4.33 to show that for every star $S \subset \mathcal{G}_{k-1}(H)$ there is a unique star containing $f_{k-1}(S)$. Recursively, we construct a sequence

$$f = f_k, f_{k-1}, \ldots, f_1,$$

where every f_i is an orthogonality and ortho-adjacency preserving transformation of $\mathcal{G}_i(H)$. In the case when $i \geq 2$, we have

$$f_i(S(Y)) \subset S(f_{i-1}(Y))$$

for every $Y \in \mathcal{G}_{i-1}(H)$ and

$$f_{i-1}(\mathcal{G}_{i-1}(X)) \subset \mathcal{G}_{i-1}(f_i(X))$$

for every $X \in \mathcal{G}_i(H)$. These inclusions imply that

$$f_1(\mathcal{G}_1(X)) \subset \mathcal{G}_1(f(X)) \quad \text{if } X \in \mathcal{G}_k(H). \tag{4.3}$$

Lemma 4.36 *The transformation f_1 is injective.*

Proof Let P and Q be distinct elements of $\mathcal{G}_1(H)$. We take mutually orthogonal

$$P_1, \ldots, P_{k-1} \in \mathcal{G}_1(H)$$

which are orthogonal to both P and Q. Consider the k-dimensional subspaces

$$X = P_1 + \cdots + P_{k-1} + P \quad \text{and} \quad Y = P_1 + \cdots + P_{k-1} + Q.$$

Then $f_1(P_1), \ldots, f_1(P_{k-1}), f_1(P)$ are mutually orthogonal and (4.3) implies that

$$f(X) = f_1(P_1) + \cdots + f_1(P_{k-1}) + f_1(P).$$

Similarly, $f_1(P_1), \ldots, f_1(P_{k-1}), f_1(Q)$ are mutually orthogonal and

$$f(Y) = f_1(P_1) + \cdots + f_1(P_{k-1}) + f_1(Q).$$

Since X and Y are adjacent, $f(X)$ and $f(Y)$ are adjacent, which implies that $f_1(P) \neq f_1(Q)$.   □

So, f_1 is an orthogonality preserving injective transformation of $\mathcal{G}_1(H)$ which sends lines of Π_H to subsets of lines (it maps every line $\mathcal{G}_1(X)$, $X \in \mathcal{G}_2(H)$ to a subset of the line $\mathcal{G}_1(f_2(X))$). By Corollary 2.7, f_1 is induced by a semilinear injective transformation of H. This semilinear injection transfers orthogonal vectors to orthogonal vectors, i.e. it is a scalar multiple of a linear or conjugate-linear isometry (by Proposition 4.2). The inclusion (4.3) shows that this isometry induces f.

The case (OA) *of Theorem 4.26.* As above, we suppose that dim $H > 2k > 2$. The required statement is a simple consequence of the following.

Lemma 4.37 *If f is an ortho-adjacency preserving injective transformation of $\mathcal{G}_k(H)$, then f is adjacency preserving.*

Proof If $X, Y \in \mathcal{G}_k(H)$ are adjacent, then $\dim(X + Y) = k + 1$ and we have

$$\dim(X + Y)^\perp \geq 2$$

(since dim $H > 2k > 2$). This implies the existence of orthogonal $P, Q \in \mathcal{G}_1(H)$ contained in $(X + Y)^\perp$. Consider the k-dimensional subspaces

$$X' = (X \cap Y) + P \quad \text{and} \quad Y' = (X \cap Y) + Q.$$

Then X, X', Y' are mutually ortho-adjacent and the same holds for Y, X', Y'. Let $\mathcal{X}$ be a maximal compatible subset of the star $\mathcal{S}(X \cap Y)$ containing X, X', Y'. Then $f(\mathcal{X})$ is a compatible subset in a star or a top. Since dim $H > 2k$, Lemma 4.30 implies that $f(\mathcal{X})$ cannot be contained in a top. Therefore, $f(\mathcal{X})$ is a subset of a star. This means that $f(\mathcal{X})$ contains the $(k - 1)$-dimensional subspace $f(X') \cap f(Y')$. Similarly, we establish that $f(X') \cap f(Y')$ is contained in $f(Y)$. Since f is injective, we have $f(X) \neq f(Y)$ and $f(X), f(Y)$ are adjacent. $\square$

Proof of Proposition 4.28 Suppose that dim $H = 2k$. Let f be an orthogonality preserving transformation of $\mathcal{G}_k(H)$ which also preserves the adjacency relation in both directions. Then for any two distinct maximal cliques $\mathcal{X}$ and $\mathcal{Y}$ of $\Gamma_k(H)$ the images $f(\mathcal{X})$ and $f(\mathcal{Y})$ are contained in distinct maximal cliques of $\Gamma_k(H)$.

Suppose that $\mathcal{X}$ is a maximal clique of $\Gamma_k(H)$ such that $f(\mathcal{X})$ is contained in two distinct maximal cliques of $\Gamma_k(H)$. One of these cliques is a star $\mathcal{S}$ and the other is a top $\mathcal{T}$ (since the intersection of two distinct maximal cliques of the same type contains at most one element). Therefore, $f(\mathcal{X})$ is contained in the line $\mathcal{S} \cap \mathcal{T}$ (a line of $\mathcal{G}_k(H)$ is a non-empty intersection of a star and a top, see Section 2.3). We take any maximal clique $\mathcal{Y}$ intersecting $\mathcal{X}$ precisely in a line and consider a maximal clique $\mathcal{Y}'$ containing $f(\mathcal{Y})$. The intersection of $f(\mathcal{X})$ and $\mathcal{Y}'$ contains more than one element. Since $f(\mathcal{X})$ is contained in the line $\mathcal{S} \cap \mathcal{T}$, the intersection of $\mathcal{Y}'$ and this line contains more than one element. This is possible only in the case when $\mathcal{Y}'$ is the star $\mathcal{S}$ or the top $\mathcal{T}$. In each of these cases, the maximal clique $\mathcal{Y}'$ contains $f(X)$ and $f(\mathcal{Y})$, which is impossible.

So, the image of every maximal clique of $\Gamma_k(H)$ is contained in precisely one maximal clique of $\Gamma_k(H)$. As in the proof of Theorem 2.15, we show that one of the following possibilities is realized:

(S) all stars go to subsets of stars,

(T) all stars go to subsets of tops.

In the case (S), we use the arguments from the proof of Theorem 4.26 and establish that f can be uniquely extended to an automorphism of $\mathcal{L}(H)$ induced by a unitary or anti-unitary operator.

In the case (T), we consider the composition of f and the orthocomplementation. This transformation satisfies (S), i.e. it is uniquely extendable to an automorphism of $\mathcal{L}(H)$ induced by a unitary or anti-unitary operator.   □

4.7 Proof of Theorem 4.29

Let f be a transformation of $\mathcal{G}_k(H)$ preserving the orthogonality relation in both directions. We suppose that $\dim H = n$ is finite and greater than $2k$.

Lemma 4.38 *The transformation f is injective.*

Proof For any distinct $X, Y \in \mathcal{G}_k(H)$ we take $Z \in \mathcal{G}_k(H)$ orthogonal to X and non-orthogonal to Y. Then $f(Z)$ is orthogonal to $f(X)$ and non-orthogonal to $f(Y)$. This implies that $f(X)$ and $f(Y)$ are distinct.   □

For $k = 1$ the statement is proved in Remark 4.9. Suppose that $k > 1$.

Lemma 4.39 *Let $X_1, \ldots X_i, Y$ be mutually distinct elements of $\mathcal{G}_k(H)$ such that Y is not contained in $X_1 + \cdots + X_i$ and*

$$\dim(X_1 + \cdots + X_i) \le n - k. \tag{4.4}$$

Then $f(Y)$ is not contained in $f(X_1) + \cdots + f(X_i)$.

Proof The condition (4.4) implies the existence of k-dimensional subspaces orthogonal to $X_1 + \cdots + X_i$. Since Y is not contained in $X_1 + \cdots + X_i$, there is $Z \in \mathcal{G}_k(H)$ orthogonal to $X_1 + \cdots + X_i$ and non-orthogonal to Y. Then $f(Z)$ is orthogonal to $f(X_1) + \cdots + f(X_i)$ and non-orthogonal to $f(Y)$.   □

By Theorem 4.26, it is sufficient to show that f is adjacency preserving. Let X and Y be adjacent elements of $\mathcal{G}_k(H)$. Consider a sequence

$$X_0, X_1, \ldots, X_{n-2k}$$

of elements from $\mathcal{G}_k(H)$ such that $X_0 = X$, $X_1 = Y$ and

$$\dim(X_0 + X_1 + \cdots + X_j) = k + j$$

for every $j \in \{1, \ldots, n - 2k\}$. Then X_j is not contained in $X_0 + \cdots + X_{j-1}$. We have $k + j \le n - k$ for every $j \in \{1, \ldots, n - 2k\}$ and Lemma 4.39 implies that

$$f(X_j) \not\subset f(X_0) + \cdots + f(X_{j-1}).$$

Therefore,

$$\dim(f(X_0) + f(X_1) + \cdots + f(X_{n-2k})) \ge \dim(f(X_0) + f(X_1)) + n - 2k - 1. \quad (4.5)$$

Since f is injective, $f(X) = f(X_0)$ and $f(Y) = f(X_1)$ are distinct. If these k-dimensional subspaces are not adjacent, then

$$\dim(f(X_0) + f(X_1)) > k + 1$$

and (4.5) shows that

$$\dim(f(X_0) + f(X_1) + \cdots + f(X_{n-2k})) > n - k.$$

The latter inequality shows that there is no k-dimensional subspace orthogonal to all $f(X_j)$. On the other hand, the equality

$$\dim(X_0 + X_1 + \cdots + X_{n-2k}) = n - k$$

implies the existence of $Z \in \mathcal{G}_k(H)$ orthogonal to all X_j. Then $f(Z)$ is orthogonal to every $f(X_j)$. This contradiction shows that $f(X)$ and $f(Y)$ are adjacent.

4.8 Proof of Theorem 4.25

Denote by $\mathcal{F}_s(H)$ the real vector space formed by all finite-rank self-adjoint operators on H. Recall that the projection on a closed subspace X is denoted by P_X and we write $\mathcal{P}_k(H)$ for the set of all rank-k projections. By Proposition 1.20, $\mathcal{F}_s(H)$ is spanned by $\mathcal{P}_k(H)$, i.e. every finite-rank self-adjoint operator can be presented as a real linear combination of some projections of rank k, for each positive integer $k < \dim H$.

Let f be a transformation of $\mathcal{G}_k(H)$ such that

$$\mathrm{tr}(f(X), f(Y)) = \mathrm{tr}(X, Y)$$

for all $X, Y \in \mathcal{G}_k(H)$. It was noted above that $\mathrm{tr}(X, Y)$ is the trace of the operators $P_X P_Y P_X$ and $P_Y P_X P_Y$, but it is also the trace of $P_X P_Y$ and $P_Y P_X$. We consider f as a transformation of $\mathcal{P}_k(H)$ which preserves the trace of the composition of any two projections.

Lemma 4.40 (Molnár [36])   *There is a unique real linear extension*

$$F : \mathcal{F}_s(H) \rightarrow \mathcal{F}_s(H)$$

of f. The transformation F is injective and

$$\mathrm{tr}(F(A)F(B)) = \mathrm{tr}(AB)$$

for any $A, B \in \mathcal{F}_s(H)$.

Proof   If $P_1, \ldots, P_m \in \mathcal{P}_k(H)$ and

$$A = a_1 P_1 + \cdots + a_m P_m \in \mathcal{F}_s(H),$$

then we define

$$F(A) = a_1 f(P_1) + \cdots + a_m f(P_m).$$

We need to show that F is well defined, i.e. if

$$A = b_1 Q_1 + \cdots + b_t Q_t$$

for some $Q_1, \ldots, Q_t \in \mathcal{P}_k(H)$, then

$$a_1 f(P_1) + \cdots + a_m f(P_m) = b_1 f(Q_1) + \cdots + b_t f(Q_t).$$

Let R be a projection of rank k. Then

$$\left(\sum_{i=1}^{m} a_i P_i - \sum_{j=1}^{t} b_j Q_j \right) R = 0.$$

Since the trace function is linear and f preserves the trace of the composition of any pair of rank-k projections, we have

$$\mathrm{tr}\left(\left(\sum_{i=1}^{m} a_i f(P_i) - \sum_{j=1}^{t} b_j f(Q_j) \right) f(R) \right) = 0.$$

In this equality, $f(R)$ can be replaced by any real linear combination

$$c_1 f(R_1) + \cdots + c_p f(R_p)$$

where $R_1, \ldots, R_p$ are projections of rank k. Therefore,

$$\mathrm{tr}\left(\left(\sum_{i=1}^{m} a_i f(P_i) - \sum_{j=1}^{t} b_j f(Q_j) \right) \left(\sum_{i=1}^{m} a_i f(P_i) - \sum_{j=1}^{t} b_j f(Q_j) \right) \right) = 0.$$

The operator

$$B = \left(\sum_{i=1}^{m} a_i f(P_i) - \sum_{j=1}^{t} b_j f(Q_j) \right)^2$$

is positive as the square of a self-adjoint operator. The trace of this operator is zero, which implies that $B = 0$ and we get the required equality.

So, F is a real linear extension of f on $\mathcal{F}_s(H)$. It is clear that this extension is unique. The linearity of the trace function guarantees that F preserves the trace of the composition of any two elements from $\mathcal{F}_s(H)$. The latter implies that F is injective. □

For any $X, Y \in \mathcal{G}_k(H)$ we denote by $\mathcal{X}_k(X, Y)$ the set of all $Z \in \mathcal{G}_k(H)$ such that $P_X + P_Y - P_Z$ is a rank-k projection.

Example 4.41 If $X, Y \in \mathcal{G}_k(H)$ are orthogonal, then $P_X + P_Y = P_{X+Y}$ is a projection of rank $2k$. For every k-dimensional subspace $Z \subset X + Y$ we have $P_{X+Y} = P_Z + P_{Z'}$, where Z' is the k-dimensional subspace of $X + Y$ orthogonal to Z. Therefore, $\mathcal{X}_k(X, Y)$ consists of all k-dimensional subspaces of $X + Y$.

Lemma 4.42 *If X and Y are closed subspaces of H, then*

$$\mathrm{Im}(P_X + P_Y) = X + Y.$$

Proof The kernel of a bounded self-adjoint operator is the orthogonal complement of the image. Also, $\mathrm{Ker}(P_X + P_Y)$ is the intersection of $\mathrm{Ker}(P_X)$ and $\mathrm{Ker}(P_Y)$. Since the operator $P_X + P_Y$ is self-adjoint, we have

$$\mathrm{Im}(P_X + P_Y) = \mathrm{Ker}(P_X + P_Y)^{\perp} = (\mathrm{Ker}(P_X) \cap \mathrm{Ker}(P_Y))^{\perp}$$
$$= (X^{\perp} \cap Y^{\perp})^{\perp} = X + Y$$

and get the claim. □

Lemma 4.42 shows that every element of $\mathcal{X}_k(X, Y)$ is contained in $X + Y$.

Lemma 4.43 *Let $X, Y \in \mathcal{G}_k(H)$, $\dim(X \cap Y) = m$ and*

$$X' = (X \cap Y)^{\perp} \cap X, \quad Y' = (X \cap Y)^{\perp} \cap Y.$$

Then $\mathcal{X}_k(X, Y)$ consists of all $(X \cap Y) + Z'$ such that $Z' \in \mathcal{X}_{k-m}(X', Y')$.

Proof First of all, we observe that

$$P_X + P_Y = P_{X'} + P_{Y'} + 2P_{X \cap Y}.$$

If $Z = (X \cap Y) + Z'$ and $Z' \in \mathcal{X}_{k-m}(X', Y')$, then $Z' \subset X' + Y'$ by Lemma 4.42, which implies that Z' is orthogonal to $X \cap Y$ and

$$P_Z = P_{Z'} + P_{X \cap Y}.$$

Since $Z' \in \mathcal{X}_{k-m}(X', Y')$, there is a $(k-m)$-dimensional subspace $M' \subset X' + Y'$ such that

$$P_{X'} + P_{Y'} = P_{Z'} + P_{M'}.$$

Note that M' is orthogonal to $X \cap Y$ and

$$P_{M'} + P_{X \cap Y} = P_M,$$

where $M = (X \cap Y) + M'$ is k-dimensional. Then

$$P_Z + P_M = P_{Z'} + P_{M'} + 2P_{X \cap Y} = P_{X'} + P_{Y'} + 2P_{X \cap Y} = P_X + P_Y$$

and Z belongs to $\mathcal{X}_k(X, Y)$.

If $Z \in \mathcal{X}_k(X, Y)$, then there is a k-dimensional subspace $M \subset X + Y$ such that

$$P_X + P_Y = P_Z + P_M$$

(it is clear that M belongs to $\mathcal{X}_k(X, Y)$). The restriction of $P_Z + P_M$ to $X \cap Y$ is $2 \operatorname{Id}_{X \cap Y}$, which implies that $X \cap Y$ is contained in $Z \cap M$. Similarly, we establish that $Z \cap M$ is contained in $X \cap Y$. Therefore, $X \cap Y = Z \cap M$ and

$$P_Z + P_M = P_{Z'} + P_{M'} + 2P_{X \cap Y},$$

where

$$Z' = (X \cap Y)^\perp \cap Z, \quad M' = (X \cap Y)^\perp \cap M.$$

Then

$$P_{X'} + P_{Y'} = P_{Z'} + P_{M'}$$

and Z' belongs to $\mathcal{X}_{k-m}(X', Y')$. $\square$

Lemma 4.44 (Gehér [25])   *The set $\mathcal{X}_k(X, Y)$ is a 1-dimensional real manifold if and only if X and Y are non-compatible and adjacent.*

Proof   It is clear that $\mathcal{X}_k(X, X) = \{X\}$ for every $X \in \mathcal{G}_k(H)$. Let X and Y be distinct k-dimensional subspaces of H. As in the proof of Proposition 4.21, we consider 2-dimensional subspaces

$$S_1, \ldots, S_m, \quad m = k - \dim(X \cap Y)$$

satisfying the following conditions:

- the subspaces $X \cap Y, S_1, \ldots, S_m$ are mutually orthogonal and their sum co-incides with $X + Y$,
- the subspaces $X_i = X \cap S_i$ and $Y_i = Y \cap S_i$ are 1-dimensional for each i.

Then

$$P_X = P_{X_1} + \cdots + P_{X_m} + P_{X \cap Y} \text{ and } P_Y = P_{Y_1} + \cdots + P_{Y_m} + P_{X \cap Y}.$$

Consider the case when X and Y are adjacent, i.e. $m = 1$. By Lemma 4.43, $Z \in \mathcal{G}_k(H)$ belongs to $X_k(X, Y)$ if and only if $Z = X' + (X \cap Y)$, where X' is an element of $X_1(X_1, Y_1)$. If X and Y are ortho-adjacent, then X_1, Y_1 are orthogonal and $X_1(X_1, Y_1) = \mathcal{G}_1(S_1)$, which implies that $X_k(X, Y)$ is a 2-dimensional real manifold. Suppose that X_1 and Y_1 are not orthogonal. Every self-adjoint operator whose image is contained in S_1 can be identified with a (2×2) - Hermitian matrix. We have

$$\mathrm{tr}(P_{X_1} + P_{Y_1}) = \mathrm{tr}(P_{X_1}) + \mathrm{tr}(P_{Y_1}) = 2$$

and the rank of $P_{X_1} + P_{Y_1}$ is two. Then

$$M = \begin{pmatrix} s & 0 \\ 0 & 2 - s \end{pmatrix}, \text{ where } 0 < s < 1,$$

is the matrix of $P_{X_1} + P_{Y_1}$ with respect to a basis formed by eigenvectors of this operator. Let Q be a 1-dimensional subspace of S_1 and let

$$N = \begin{pmatrix} a_1 & b \\ \bar{b} & a_2 \end{pmatrix}$$

be the Hermitian matrix corresponding to the projection P_Q with respect to the same basis. Since

$$\mathrm{tr}(M - N) = \mathrm{tr}(M) - \mathrm{tr}(N) = 2 - 1 = 1,$$

we have $Q \in X_1(X_1, Y_1)$ if and only if $\mathrm{rank}(M - N) = 1$, which is equivalent to

$$\mathrm{rank}(I_2 - M^{-1}N) = 1.$$

Since $\mathrm{rank}(M^{-1}N) = 1$, the latter holds if and only if $\mathrm{tr}(M^{-1}N) = 1$. So, Q belongs to $X_1(X_1, Y_1)$ if and only if

$$\mathrm{tr}(N) = a_1 + a_2 = 1, \quad \mathrm{tr}(M^{-1}N) = a_1/s + a_2/(2 - s) = 1, \quad \mathrm{rank}(N) = 1.$$

The first and second equalities imply that $a_1 = s/2$ and $a_2 = (2 - s)/2$. Using the third condition, we establish that

$$b = \frac{\sqrt{s(2 - s)}}{2} e^{it} \text{ with } t \in [0, 2\pi).$$

This means that $X_1(X_1, Y_1)$ is a 1-dimensional real manifold and the same holds for $X_k(X, Y)$.

If $m \geq 2$, then $X_k(X, Y)$ contains any k-dimensional subspace of type

$$X'_1 + \cdots + X'_m + X \cap Y,$$

where each X'_i belongs to $X_1(X_i, Y_i)$. The above arguments show that $X_k(X, Y)$ contains (as a subset) a real manifold of dimension at least two, i.e. it cannot be a 1-dimensional real manifold. $\square$

Now, we prove Theorem 4.25. It was noted above that we can restrict ourselves to the case when $\dim H \geq 2k$. First of all, we observe that f is orthogonality preserving in both directions (indeed, $X, Y \in G_k(H)$ are orthogonal if and only if $\operatorname{tr}(X, Y) = 0$). Since f (as a transformation of $\mathcal{P}_k(H)$) is extendable to the linear injection F of $\mathcal{F}_s(H)$ to itself, we have

$$f(X_k(X, Y)) \subset X_k(f(X), f(Y))$$

for any $X, Y \in G_k(H)$. If X and Y are orthogonal elements of $G_k(H)$, then $f(X)$ and $f(Y)$ are orthogonal. In this case, $X_k(X, Y)$ and $X_k(f(X), f(Y))$ consist of all k-dimensional subspaces contained in $X + Y$ and $f(X) + f(Y)$ (respectively), which implies that

$$f(G_k(X + Y)) \subset G_k(f(X) + f(Y)).$$

Suppose that $k = 1$. This case was considered in Section 4.3, but we want to show that the required statement follows from Corollary 2.7 (under the assumption that $\dim H \geq 3$). Every 2-dimensional subspace $S \subset H$ can be presented as the sum of orthogonal 1-dimensional subspaces $P, Q \subset S$. Then f transfers every 1-dimensional subspace of S to a 1-dimensional subspace of the 2-dimensional subspace $f(P) + f(Q)$. Therefore, f sends lines of Π_H to subsets of lines. Also, f is injective and its image contains three mutually orthogonal 1-dimensional subspaces. So, f satisfies the conditions of Corollary 2.7, i.e. it is induced by a semilinear injection of H to itself. By Proposition 4.2, this semilinear injection is a scalar multiple of a linear or conjugate-linear isometry.

To prove Theorem 4.25 for $k \geq 2$ we need the following.

Lemma 4.45 *If M is a connected compact topological real manifold, then every injective continuous map $h : M \to M$ is bijective, i.e. it is a homeomorphism of M to itself.*

Proof Suppose that M is m-dimensional. For every $a \in h(M)$ consider an open subset $U_a \subset M$ homeomorphic to $\mathbb{R}^m$ and containing a. There exists an open subset $W_a \subset M$ homeomorphic to $\mathbb{R}^m$, containing $h^{-1}(a)$ and such that $h(W_a) \subset U_a$. By [16, Chapter IV, Proposition 7.4], $h(W_a)$ is an open subset of

U_a and, consequently, $h(W_a)$ is an open subset of M. Since $h(W_a) \subset h(M)$ for every $a \in h(M)$, the image $h(M)$ is an open subset of M. On the other hand, $h(M)$ is compact, i.e. it is an open–closed subset of M and we have $h(M) = M$ by the connectedness of M. □

Let $\dim H \geq 2k > 2$. Suppose that $\dim H = n$ is finite. Then the vector space $\mathcal{F}_s(H)$ is finite-dimensional and F is a linear automorphism. Also, $\mathcal{G}_k(H)$ is a connected compact topological real manifold of dimension $2k(n - k)$ whose topology coincides with the topology defined by the gap metric. Then f is a continuous injection of $\mathcal{G}_k(H)$ to itself (as the restriction of F to $\mathcal{G}_k(H)$). By Lemma 4.45,

$$f(\mathcal{G}_k(H)) = \mathcal{G}_k(H)$$

and f is a homeomorphism of $\mathcal{G}_k(H)$ to itself. Since F is a linear automorphism of $\mathcal{F}_s(H)$, for any $X, Y \in \mathcal{G}_k(H)$ we have

$$f(\mathcal{X}_k(X, Y)) = \mathcal{X}_k(f(X), f(Y)).$$

By Lemma 4.44, the latter equality together with the fact that f is a homeomorphism imply that $X, Y \in \mathcal{G}_k(H)$ are non-compatible and adjacent if and only if the same holds for $f(X), f(Y)$. The set of all elements of $\mathcal{G}_k(H)$ adjacent to $X \in \mathcal{G}_k(H)$ is the closure of the set of all $Y \in \mathcal{G}_k(H)$ such that X, Y are non-compatible and adjacent. So, f is adjacency preserving in both directions. Chow's theorem (Theorem 2.15) together with Proposition 4.2 give the claim.

Now, we assume that H is infinite-dimensional. If $X, Y \in \mathcal{G}_k(H)$ are orthogonal, then $f(X), f(Y)$ are orthogonal and

$$f(\mathcal{G}_k(X + Y)) \subset \mathcal{G}_k(f(X) + f(Y)).$$

We apply the above arguments to the restriction of f to $\mathcal{G}_k(X + Y)$. For any $X', Y' \in \mathcal{G}_k(H)$ there are orthogonal $X, Y \in \mathcal{G}_k(H)$ such that $X + Y$ contains X' and Y'. This implies that f is adjacency preserving in both directions. Since f is also orthogonality preserving, we can use Theorem 4.26.

Remark 4.46 Theorem 4.26 is not exploited in the original proof of Theorem 4.25. In the infinite-dimensional case, the statement can be proved as follows. For every $2k$-dimensional subspace $X \subset H$ there is a $2k$-dimensional subspace $X' \subset H$ and a unitary or anti-unitary operator $U_X : X \to X'$ such that the restriction of f to $\mathcal{G}_k(X)$ is induced by U_X or it sends every $Y \in \mathcal{G}_k(X)$ to $U_X(Y)^{\perp} \cap X'$. First, we need to show that the second possibility is not realized. After that, we construct a unitary or anti-unitary operator U on H whose restriction to any $2k$-dimensional subspace $X \subset H$ is a scalar multiple of U_X.

4.9 Transformations of Grassmannians Preserving the Gap Metric

Recall that the *gap* between two closed subspaces $X, Y \subset H$ is defined as

$$g(X, Y) = \|P_X - P_Y\|.$$

Since $P_{X^\perp} - P_{Y^\perp} = (\mathrm{Id}_H - P_X) - (\mathrm{Id}_H - P_Y) = P_Y - P_X$, we obtain that

$$g(X, Y) = g(X^\perp, Y^\perp).$$

In this section, we consider the Grassmannian $\mathcal{G}_k(H)$ as the metric space, where the distance between $X, Y \in \mathcal{G}_k(H)$ is equal to $g(X, Y)$. By Remark 4.19, the gap $g(X, Y)$ is equal to the sine of the largest principal angle between X and Y.

The following statement describes all bijective isometries of $\mathcal{G}_k(H)$ with respect to the gap metric.

Theorem 4.47 (Botelho, Jamison and Molnár [9] and Gehér and Šemrl [24]) *Every bijective transformation of $\mathcal{G}_k(H)$ preserving the gap metric can be uniquely extended to an automorphism of the lattice $\mathcal{L}(H)$ induced by a unitary or anti-unitary operator on H or $\dim H = 2k$ and it is uniquely extendable to an anti-automorphism of $\mathcal{L}(H)$ induced by a unitary or anti-unitary operator.*

The proof follows at the end of this section.

Remark 4.48   Theorem 4.47 was first proved in [9] for the case when $\dim H \geq 4k$. The general case was considered in [24].

Remark 4.49   If H is finite-dimensional, then $\mathcal{G}_k(H)$ is a connected compact topological manifold and every transformation of $\mathcal{G}_k(H)$ preserving the gap metric is a continuous injection and, consequently, it is bijective by Lemma 4.45.

The above statement follows from Wigner's theorem if $k = 1$. If $\dim H = n$ is finite, then the orthocomplementation map of $\mathcal{G}_k(H)$ to $\mathcal{G}_{n-k}(H)$ is an isometry with respect to the gap metric. Therefore, it is sufficient to prove Theorem 4.47 only for the case when $\dim H \geq 2k > 2$.

For any two $X, Y \in \mathcal{G}_k(H)$ we denote by $\mathcal{M}(X, Y)$ the set of all $Z \in \mathcal{G}_k(H)$ such that

$$g(X, Z) \leq 1/\sqrt{2} \quad \text{and} \quad g(Y, Z) \leq 1/\sqrt{2}.$$

The equality $g(X, Y) = 1$ implies that the largest principal angle between X and Y is $\pi/2$; in particular, it holds if X, Y are orthogonal.

Lemma 4.50 (Gehér and Šemrl [24])   *If $\dim H \geq 2k$, then for any $X, Y \in \mathcal{G}_k(H)$ satisfying $g(X, Y) = 1$ the following two conditions are equivalent:*

- *X and Y are orthogonal,*
- *$M(X, Y)$ is a compact manifold.*

The proof of Lemma 4.50 is very complicated and we do not give it here. Let f be a bijective transformation of $\mathcal{G}_k(H)$ preserving the gap metric. Then

$$f(M(X, Y)) = M(f(X), f(Y))$$

for all $X, Y \in \mathcal{G}_k(H)$ and Lemma 4.50 shows that f is orthogonality preserving in both directions. Theorem 4.10 gives the claim if $\dim H > 2k$.

In the case when $\dim H = 2k$, there is the following characterization of the complementary relation in terms of the gap metric.

Lemma 4.51 *Suppose that $\dim H = 2k$. Then $X, Y \in \mathcal{G}_k(H)$ are complementary if and only if $g(X^{\perp}, Y) < 1$.*

Proof If $X + Y$ is a proper subspace of H, then there is a 1-dimensional subspace orthogonal to both X and Y. Since this 1-dimensional subspace is contained in $X^{\perp}$, the largest principal angle between $X^{\perp}$ and Y is $\pi/2$, which means that $g(X^{\perp}, Y) = 1$.

In the case when $X + Y = H$, we consider the mutually orthogonal 2-dimensional subspaces $S_1, \ldots, S_k$ defined as in the proof of Proposition 4.21. Each S_i intersects X and Y in distinct 1-dimensional subspaces which will be denoted by X_i and Y_i, respectively. Then

$$X^{\perp} = X_1' + \cdots + X_k',$$

where X_i' is the orthogonal complement of X_i in S_i. For every i the angle $\angle(X_i', Y_i)$ is one of the principal angles between $X^{\perp}$ and Y. Since $X_i \neq Y_i$, each $\angle(X_i', Y_i)$ is less than $\pi/2$. Therefore, $g(X^{\perp}, Y) < 1$. $\qquad\square$

Proof of Theorem 4.47 for $\dim H = 2k$ Since f is orthogonality preserving in both directions, we have

$$f(X^{\perp}) = f(X)^{\perp}$$

for all $X \in \mathcal{G}_k(H)$. Then

$$g(X^{\perp}, Y) = g(f(X^{\perp}), f(Y)) = g(f(X)^{\perp}, f(Y))$$

for any $X, Y \in \mathcal{G}_k(H)$ and Lemma 4.51 shows that f is complementary preserving in both directions. By Theorems 2.15 and 2.21, f can be extended to an automorphism or anti-automorphism of $\mathcal{L}(H)$. The statement follows from the fact that f is orthogonality preserving. $\qquad\square$

5

Compatibility Relation

In this chapter, we investigate compatibility preserving transformations of the lattice of closed subspaces of a complex Hilbert space and the associated Grassmannians. Since two 1-dimensional subspaces are compatible if and only if they are orthogonal, the results of this chapter can be considered as Wigner type theorems.

In almost all cases, we suppose that a transformation is bijective and preserves the compatibility relation in both directions. We apply combinatorial arguments in the spirit of Mackey [35] and Rickart [54] to transformations of the lattice and the Grassmannian formed by closed subspaces of infinite dimension and codimension. If the Grassmannian consists of finite-dimensional subspaces, then we use orthogonal apartments, i.e. maximal compatible subsets of this Grassmannian, and their maximal intersections (a modification of the idea exploited in Section 2.6). It must be pointed out that compatibility preserving transformations are not determined in some low-dimensional cases and we have a few open problems.

5.1 Compatibility Preserving Transformations

Let H be a complex Hilbert space whose dimension is not less than three. For closed subspaces $X, Y \subset H$ the following three conditions are equivalent:

- $(X \cap Y)^{\perp} \cap X$ and $(X \cap Y)^{\perp} \cap Y$ are orthogonal, i.e. X and Y are compatible;
- there is an orthonormal basis of H such that X and Y are spanned by subsets of this basis, in other words, there is an orthogonal apartment of $\mathcal{L}(H)$ containing X and Y;
- the projections P_X and P_Y commute.

Every automorphism or anti-automorphism of the lattice $\mathcal{L}(H)$ induced by a unitary or anti-unitary operator on H preserves the compatibility relation in both directions, but there exist other bijective transformations of $\mathcal{L}(H)$ satisfying this condition.

Example 5.1 Let $\mathcal{X}$ be a subset of $\mathcal{L}(H)$ such that for every $X \in \mathcal{X}$ the orthogonal complement $X^{\perp}$ also belongs to $\mathcal{X}$. Denote by $\pi_{\mathcal{X}}$ the bijective transformation of $\mathcal{L}(H)$ defined as follows:

$$\pi_{\mathcal{X}}(X) = \begin{cases} X^{\perp} & \text{if } X \in \mathcal{X}, \\ X & \text{if } X \notin \mathcal{X}. \end{cases}$$

This transformation preserves the compatibility relation in both directions (since every $Y \in \mathcal{L}(H)$ compatible to $X \in \mathcal{L}(H)$ is also compatible to $X^{\perp}$). In the case when $\mathcal{X}$ coincides with $\mathcal{L}(H)$, this transformation is the orthocomplementation.

We state that every bijective transformation of $\mathcal{L}(H)$ preserving the compatibility relation in both directions is an automorphism of the lattice $\mathcal{L}(H)$ induced by a unitary or anti-unitary operator or the composition of such an automorphism and a certain $\pi_{\mathcal{X}}$.

Theorem 5.2 (Molnár and Šemrl [40]) *Let f be a bijective transformation of $\mathcal{L}(H)$ preserving the compatibility relation in both directions, i.e. $X, Y \in \mathcal{L}(H)$ are compatible if and only if $f(X), f(Y)$ are compatible. Then there exists an automorphism g of the lattice $\mathcal{L}(H)$ induced by a unitary or anti-unitary operator and such that for every $X \in \mathcal{L}(H)$ we have either*

$$f(X) = g(X) \quad or \quad f(X) = g(X)^{\perp}. \tag{5.1}$$

In the case when H is infinite-dimensional, the same holds for the Grassmannian $\mathcal{G}_{\infty}(H)$ formed by all closed subspaces of H whose dimension and codimension both are infinite.

Theorem 5.3 (Plevnik [51]) *Suppose that H is infinite-dimensional and f is a bijective transformation of $\mathcal{G}_{\infty}(H)$ preserving the compatibility relation in both directions. Then there exists an automorphism g of the lattice $\mathcal{L}(H)$ induced by a unitary or anti-unitary operator and such that for every $X \in \mathcal{G}_{\infty}(H)$ we have (5.1).*

The proofs of Theorems 5.2 and 5.3 are given in Section 5.2.

Remark 5.4 Recall that $\mathcal{I}_{\infty}(H)$ is the set of all idempotents in the algebra $\mathcal{B}(H)$ whose image and kernel both are infinite-dimensional (Remark 3.22). In [51], Plevnik describes commutativity preserving bijective transformations of $\mathcal{I}_{\infty}(H)$. We use the same arguments to prove Theorem 5.3.

Theorem 5.5 (Pankov [46]) *If H is infinite-dimensional, then for each natural k every bijective transformation of $\mathcal{G}_k(H)$ preserving the compatibility relation in both directions can be uniquely extended to an automorphism of the lattice $\mathcal{L}(H)$ induced by a unitary or anti-unitary operator.*

The proof of Theorem 5.5 is given in Section 5.3.

For $k = 1$ this statement follows immediately from Proposition 4.8, since two distinct elements of $\mathcal{G}_1(H)$ are compatible if and only if they are orthogonal. In the case when $k > 1$, we will exploit the notion of orthogonal apartment (in the next section, we explain why the arguments from the proof of Theorems 5.2 and 5.3 do not work). Recall that for every orthonormal basis B of H the associated *orthogonal apartment* of $\mathcal{G}_k(H)$ consists of all k-dimensional subspaces spanned by subsets of B. By Proposition 1.15, orthogonal apartments can be characterized as maximal compatible subsets of $\mathcal{G}_k(H)$ and every compatible subset of $\mathcal{G}_k(H)$ is contained in an orthogonal apartment. Therefore, for a bijective transformation f of $\mathcal{G}_k(H)$ the following two conditions are equivalent:

- f and f^{-1} send orthogonal apartments to orthogonal apartments,
- f preserves the compatibility relation in both directions.

As in the proof of Theorem 2.24 (Section 2.6), we determine maximal intersections of two orthogonal apartments. Using such intersections, we show that every bijective transformation f of $\mathcal{G}_k(H)$ satisfying the above conditions is orthogonality preserving in both directions if H is infinite-dimensional.

Now, we consider the case when H is finite-dimensional.

Theorem 5.6 (Pankov [46]) *Let f be a compatibility preserving bijective transformation of $\mathcal{G}_k(H)$, i.e. f sends any pair of compatible elements to compatible elements. The dimension of H is assumed to be finite and distinct from $2k$, and we require that k is distinct from 2 and 4 if $\dim H = 6$. Then f can be uniquely extended to an automorphism of the lattice $\mathcal{L}(H)$ induced by a unitary or anti-unitary operator.*

Applying Theorem 4.26, we prove the same for compatibility preserving injections under some additional assumptions.

Theorem 5.7 *Let f be a compatibility preserving injective transformation of $\mathcal{G}_k(H)$. We suppose that the dimension of H is finite and distinct from $2k$. Suppose also that the following possibility is not realized:*

- $\dim H = n$ is even and $(n - 2k)^2 = n - 2$.

Then f is a bijective transformation which can be uniquely extended to an automorphism of the lattice $\mathcal{L}(H)$ induced by a unitary or anti-unitary operator.

Remark 5.8 The equality $(n - 2k)^2 = n - 2$ holds, for example, if $n = 6$ and $k = 2, 4$ or $n = 102$ and $k = 46, 56$.

The proofs of Theorems 5.6 and 5.7 are given in Section 5.5.

For $k = 1$ the above statements hold by Remark 4.9. If $\dim H = n$ is finite, then it is sufficient to prove Theorems 5.6 and 5.7 only for the case when $k < n - k$. Indeed, if f is a compatibility preserving transformation of $\mathcal{G}_k(H)$, then

$$X \to f(X^\perp)^\perp, \quad X \in \mathcal{G}_{n-k}(H)$$

is a compatibility preserving transformation of $\mathcal{G}_{n-k}(H)$. If the latter transformation is extendable to an automorphism of $\mathcal{L}(H)$ induced by a unitary or anti-unitary operator, then the restriction of this lattice automorphism to $\mathcal{G}_k(H)$ coincides with f.

If $\dim H = n$ is finite, then every orthogonal apartment of $\mathcal{G}_k(H)$, i.e. a maximal compatible subset of $\mathcal{G}_k(H)$, consists of $\binom{n}{k}$ elements. In this case, for an injective transformation f of $\mathcal{G}_k(H)$ the following two conditions are equivalent:

(1) f sends orthogonal apartments to orthogonal apartments,
(2) f is compatibility preserving.

The implication (1) $\Longrightarrow$ (2) is obvious. Conversely, if f is compatibility preserving and $\mathcal{A}$ is an orthogonal apartment of $\mathcal{G}_k(H)$, then $f(\mathcal{A})$ is a compatible subset consisting of $\binom{n}{k}$ elements which implies that it is an orthogonal apartment. Note that the second condition does not imply the first if H is infinite-dimensional.

To prove Theorems 5.6 and 5.7 we characterize the ortho-adjacency and orthogonality relations in terms of maximal intersections of orthogonal apartments. In the case when $n = 6$ and k is equal to 2 or 4, there is no such characterization for the ortho-adjacency. Similarly, we cannot characterize the orthogonality relation if the possibility mentioned in Theorem 5.7 is realized. For this reason, we need the corresponding assumptions in Theorems 5.6 and 5.7.

In the case when $\dim H = 2k$, we obtain the following result similar to Theorems 5.2 and 5.3.

Theorem 5.9 (Pankov [46]) *Suppose that $\dim H = 2k \geq 8$ and f is a bijective transformation of $\mathcal{G}_k(H)$ preserving the compatibility relation in both directions. Then there exists an automorphism g of the lattice $\mathcal{L}(H)$ induced by* *a unitary or anti-unitary operator and such that for every $X \in \mathcal{G}_k(H)$ we have either*

$$f(X) = g(X) \ \ or \ \ f(X) = g(X)^{\perp}.$$

The proof of Theorem 5.9 is given in Section 5.6.

5.2 Proofs of Theorems 5.2 and 5.3

For every subset $X \subset \mathcal{L}(H)$ we denote by X^c the set of all elements of $\mathcal{L}(H)$ compatible to every element of X. Note that 0 and H always belong to X^c. We write X^{cc} for $\mathcal{Y}^c$, where $\mathcal{Y} = X^c$.

Let X and Y be distinct compatible elements of $\mathcal{L}(H)$ both different from 0 and H. Then H can be presented as the sum of the following mutually orthogonal subspaces:

$$Z_1 = X \cap Y, \quad Z_2 = X^{\perp} \cap Y, \quad Z_3 = X \cap Y^{\perp}, \quad Z_4 = X^{\perp} \cap Y^{\perp}$$

(some of which may be zero) and $\{X, Y\}^{cc}$ consists of all sums

$$\sum_{i \in I} Z_i \ \text{with} \ I \subset \{1, 2, 3, 4\}$$

(note that 0 and H correspond to the cases when $I = \emptyset$ and $I = \{1, 2, 3, 4\}$, respectively). Some of these sums may be coincident and we have

$$|\{X, Y\}^{cc}| = 2^k,$$

where k is the number of non-zero Z_i.

Using the number of elements in the set $\{X, Y\}^{cc}$, we can distinguish elements from $\mathcal{G}_1(H) \cup \mathcal{G}^1(H)$.

Lemma 5.10 *If $X \in \mathcal{G}_1(H) \cup \mathcal{G}^1(H)$, then for every $Y \in \mathcal{L}(H) \setminus \{0, H\}$ compatible to X we have*

$$|\{X, Y\}^{cc}| \in \{4, 8\}.$$

In the case when $X \in \mathcal{L}(H) \setminus \{0, H\}$ does not belong to $\mathcal{G}_1(H) \cup \mathcal{G}^1(H)$, there is $Y \in \mathcal{L}(H)$ compatible to X and such that

$$|\{X, Y\}^{cc}| = 16.$$

Proof Suppose that X is an element of $\mathcal{G}_1(H) \cup \mathcal{G}^1(H)$. If Y is the orthogonal complement of X, then $Z_1 = Z_4 = 0$ and $\{X, Y\}^{cc}$ consists of four elements. For all other cases, there are precisely three non-zero Z_i and the number of elements in $\{X, Y\}^{cc}$ is equal to eight.

Suppose that $X \in \mathcal{L}(H) \setminus \{0, H\}$ does not belong to $\mathcal{G}_1(H) \cup \mathcal{G}^1(H)$. We take any $Y \in \mathcal{L}(H)$ compatible to X and such that $X \cap Y$ is non-zero and $X + Y$ is a proper subspace of H. Then all Z_i are non-zero and the number of elements in $\{X, Y\}^{cc}$ is equal to 16. $\qquad\square$

To prove Theorem 5.3 we will consider the intersection of $\mathcal{G}_\infty(H)$ with $\{X, Y\}^{cc}$ for $X, Y \in \mathcal{G}_\infty(H)$.

Lemma 5.11 *Suppose that H is infinite-dimensional. If X, Y are distinct compatible elements from $\mathcal{G}_\infty(H)$ and $Y \neq X^\perp$, then the following two conditions are equivalent:*

(1) $|\{X, Y\}^{cc} \cap \mathcal{G}_\infty(H)| \in \{4, 6\}$,
(2) $X \perp Y$ or $X^\perp \perp Y$ or $X \perp Y^\perp$ or $X^\perp \perp Y^\perp$.

Proof The assumption $Y \neq X^\perp$ implies that at least three Z_i are non-zero.

The condition (2) is equivalent to the fact that one of the Z_i is zero, i.e. there are precisely three non-zero Z_i. We have

$$X = Z_1 + Z_3, \quad X^\perp = Z_2 + Z_4 \quad \text{and} \quad Y = Z_1 + Z_2, \quad Y^\perp = Z_3 + Z_4.$$

This guarantees that at most one of the non-zero Z_i is finite-dimensional (otherwise, at most one of the Z_i is infinite-dimensional, which implies that X or Y has a finite dimension or codimension and we get a contradiction). If all non-zero Z_i are infinite-dimensional, then

$$|\{X, Y\}^{cc} \cap \mathcal{G}_\infty(H)| = 6$$

(each non-zero Z_i and the sum of any two non-zero Z_i are elements of $\mathcal{G}_\infty(H)$). If one of the non-zero Z_l is finite-dimensional, then it and its orthogonal complement do not belong to $\mathcal{G}_\infty(H)$ and we have

$$|\{X, Y\}^{cc} \cap \mathcal{G}_\infty(H)| = 4.$$

If the condition (2) does not hold, then all Z_i are non-zero. In this case, there are at most two finite-dimensional Z_i (otherwise, at most one of the Z_i is infinite-dimensional and X or Y has a finite dimension or codimension). If all Z_i are infinite-dimensional, then every orthogonal sum $\sum_{i \in I} Z_i$, where I is a proper subset of $\{1, 2, 3, 4\}$, belongs to $\mathcal{G}_\infty(H)$ and

$$|\{X, Y\}^{cc} \cap \mathcal{G}_\infty(H)| = 14.$$

If only one of the Z_i is finite-dimensional, then it and its orthogonal complement do not belong to $\mathcal{G}_\infty(H)$ and we have

$$|\{X, Y\}^{cc} \cap \mathcal{G}_\infty(H)| = 12.$$

Similarly, if Z_i and Z_j are finite-dimensional, then $Z_i, Z_j, Z_i + Z_j$ and their orthogonal complements do not belong to $G_\infty(H)$, which means that

$$|\{X, Y\}^{cc} \cap G_\infty(H)| = 8.$$

We get the claim. $\qquad\qquad\qquad\qquad\qquad\qquad\qquad\qquad\qquad\qquad\qquad\square$

Remark 5.12   Ideas of this kind were first exploited in [35, 54] (see also [15, Chapter IV] or [44, Section 3.7]).

Proof of Theorem 5.2   Let f be a bijective transformation of $\mathcal{L}(H)$ preserving the compatibility relation in both directions. If a closed subspace is compatible to all elements of $\mathcal{L}(H)$, then this subspace is 0 or H. Therefore, f transfers the set $\{0, H\}$ to itself. Since for distinct $X, Y \in \mathcal{L}(H)$ we have $Y = X^\perp$ if and only if $|\{X, Y\}^{cc}| = 2$, f preserves the orthocomplementation, i.e. $f(X)^\perp = f(X^\perp)$ for all $X \in \mathcal{L}(H)$. Lemma 5.10 implies that f sends $G_1(H) \cup G^1(H)$ to itself.

Consider the transformation h of $G_1(H)$ defined for every $X \in G_1(H)$ as follows:

$$h(X) = \begin{cases} f(X) & \text{if } f(X) \in G_1(H), \\ f(X)^\perp & \text{if } f(X) \in G^1(H). \end{cases}$$

First, we show that this transformation is injective.

Suppose that $h(X) = h(Y)$ for some distinct $X, Y \in G_1(H)$. Then one of $f(X), f(Y)$ is a 1-dimensional subspace and the other is a hyperplane. Consider the case when $f(X) \in G_1(H)$ and $f(Y) \in G^1(H)$. The equality $h(X) = h(Y)$ implies that $f(Y)^\perp = f(X)$. Then $f(Y^\perp) = f(X)$ and we get a contradiction, since f is bijective and $Y^\perp \neq X$.

Now, we establish that h is surjective. Let $Y \in G_1(H)$. If $Y = f(X)$ or $Y^\perp = f(X)$ for some $X \in G_1(H)$, then $h(X) = Y$. If $Y = f(U)$ and $Y^\perp = f(V)$ for some $U, V \in G^1(H)$, then there is no $X \in G_1(H)$ such that $h(X) = Y$; but this case is not realized. Indeed, we have

$$f(V) = Y^\perp = f(U)^\perp = f(U^\perp),$$

which is impossible, since f is bijective and $V \neq U^\perp$.

So, h is bijective. Also, h preserves the orthogonality relation in both directions and, by Proposition 4.8, it can be extended to a certain automorphism g of the lattice $\mathcal{L}(H)$ induced by a unitary or anti-unitary operator. The transformation $g^{-1}f$ preserves the compatibility relation in both directions. It is easy to see that $g^{-1}f$ leaves fixed every $X \in G_1(H)$ or sends it to the orthogonal complement $X^\perp$. Therefore, $Y \in \mathcal{L}(H)$ is compatible to $X \in G_1(H)$ if and only if $g^{-1}f(Y)$ is compatible to X (any element of $\mathcal{L}(H)$ is compatible to X if and only if it is compatible to $X^\perp$). Note that a 1-dimensional subspace is compatible to Y if and only if it is contained in Y or $Y^\perp$. We need to show that $g^{-1}f(Y)$ coincides with Y or $Y^\perp$ for every $Y \in \mathcal{L}(H)$, which means that the lattice automorphism g is as required.

First of all, we observe that $g^{-1}f(Y)$ is contained in Y or $Y^\perp$ (otherwise, $g^{-1}f(Y)$ contains a 1-dimensional subspace non-compatible to Y and $Y^\perp$). If $g^{-1}f(Y)$ is a proper subspace of Y, then there is a 1-dimensional subspace of Y non-compatible to $g^{-1}f(Y)$. Similarly, we obtain that $g^{-1}f(Y) = Y^\perp$ if $g^{-1}f(Y)$ is contained in $Y^\perp$.   □

Proof of Theorem 5.3   Suppose that H is infinite-dimensional and f is a bijective transformation of $\mathcal{G}_\infty(H)$ preserving the compatibility relation in both directions. For distinct compatible elements $X, Y \in \mathcal{G}_\infty(H)$ we have

$$|\{X, Y\}^{cc} \cap \mathcal{G}_\infty(H)| = 2$$

if and only if Y is the orthogonal complement of X. Therefore, f preserves the orthocomplementation, i.e. $f(X^\perp) = f(X)^\perp$ for every $X \in \mathcal{G}_\infty(H)$.

We will construct a bijective transformation g of $\mathcal{G}_\infty(H)$ satisfying the following conditions:

* g is orthogonality preserving in both directions,
* for every $X \in \mathcal{G}_\infty(H)$ we have $g(X) = f(X)$ or $g(X) = f(X)^\perp$.

Then, by Theorem 4.6, g can be extended to a lattice automorphism induced by a unitary or anti-unitary operator and we get the claim.

Let $X \in \mathcal{G}_\infty(H)$. We take any $Y \in \mathcal{G}_\infty(H)$ orthogonal to X and distinct from $X^\perp$. By Lemma 5.11, one of the following possibilities is realized:

(1) $f(X)$ is orthogonal to $f(Y)$ or $f(Y)^\perp$,
(2) $f(X)^\perp$ is orthogonal to $f(Y)$ or $f(Y)^\perp$.

In the first case, we set $g(X) = f(X)$. In the second case, we define $g(X)$ as the orthogonal complement of $f(X)$. We need to show that the definition of $g(X)$ does not depend on the choice of element $Y \neq X^\perp$ orthogonal to X. In other words, if the possibility (i), $i \in \{1, 2\}$ is realized for a certain $Y \in \mathcal{G}_\infty(H) \setminus \{X^\perp\}$ orthogonal to X, then the same possibility is realized for all such Y.

Let Y and Z be distinct elements of $\mathcal{G}_\infty(H)$ orthogonal to X and distinct from $X^\perp$. First, we consider the case when Y and Z are non-compatible. Suppose that $f(X)$ is orthogonal to $f(Y)$ or $f(Y)^\perp$ and $f(X)^\perp$ is orthogonal to $f(Z)$ or $f(Z)^\perp$. In other words, one of $f(Y), f(Y)^\perp$ is contained in $f(X)^\perp$ and one of $f(Z), f(Z)^\perp$ is contained in $f(X)$. This means that one of $f(Y), f(Y)^\perp$ is compatible to one of $f(Z), f(Z)^\perp$. Since f is compatibility preserving in both directions, one of $Y, Y^\perp$ is compatible to one of $Z, Z^\perp$, which contradicts the fact that Y and Z are non-compatible. Therefore, $f(X)$ is orthogonal to $f(Y)$ or $f(Y)^\perp$ if and only if it is orthogonal to $f(Z)$ or $f(Z)^\perp$.

In the case when Y and Z are compatible, we choose any $Y' \in \mathcal{G}_\infty(H) \setminus \{X^\perp\}$ orthogonal to X and non-compatible to both Y and Z. We apply the above arguments to the pairs Y, Y' and Z, Y' So, the transformation g is well defined.

Since f is bijective and preserves the orthocomplementation, we have $g(X) \neq g(Y)$ for any distinct $X, Y \in \mathcal{G}_\infty(H)$ such that $Y \neq X^\perp$. It is clear that $g(X^\perp)$ coincides with $g(X)$ or $g(X)^\perp$, but we cannot state that $g(X) \neq g(X^\perp)$ at this moment. We need to show that g is bijective. Let us consider the inverse transformation f^{-1} and the associated transformation g' defined as g for f.

Let $X \in \mathcal{G}_\infty(H)$ and $X' = g(X)$. Suppose that $g(X) = f(X)$. Then for every $Y \in \mathcal{G}_\infty(H) \setminus \{X^\perp\}$ orthogonal to X we have

$$f(X) \perp f(Y) \quad \text{or} \quad f(X) \perp f(Y)^\perp.$$

If $f(X)$ is orthogonal to $f(Y)$, then we consider $Y' = f(Y)$, which is orthogonal to X' and distinct from $X'^\perp$. Since $f^{-1}(X') = X$ and $f^{-1}(Y') = Y$ are orthogonal, we have $g'(X') = X$. In the case when $f(X)$ is orthogonal to $f(Y)^\perp$, we take $Y' = f(Y)^\perp$. As above, Y' is orthogonal to X' and distinct from $X'^\perp$. Since $f^{-1}(X') = X$ is orthogonal to $f^{-1}(Y')^\perp = Y$, we get $g'(X') = X$ again.

Now, we suppose that $g(X) = f(X)^\perp$. Then for every $Y \in \mathcal{G}_\infty(H) \setminus \{X^\perp\}$ orthogonal to X we have

$$f(X)^\perp \perp f(Y) \quad \text{or} \quad f(X)^\perp \perp f(Y)^\perp.$$

Consider the first possibility (the second is similar). In this case, $Y' = f(Y)$ is orthogonal to $X' = f(X)^\perp$ and distinct from $X'^\perp$. Then $f^{-1}(X')^\perp = X$ and $f^{-1}(Y') = Y$ are orthogonal. This means that $g'(X') = X$.

So, for every $X \in \mathcal{G}_\infty(H)$ we have $g'g(X) = X$ and the same arguments show that $gg'(X) = X$. Therefore, g is bijective, which guarantees that

$$g(X^\perp) = g(X)^\perp$$

for every $X \in \mathcal{G}_\infty(H)$. To complete the proof we need to establish that g is orthogonality preserving in both directions.

Suppose that $X, Y \in \mathcal{G}_\infty(H)$ are orthogonal and $Y \neq X^\perp$. If $g(X) = f(X)$, then $f(X)$ is orthogonal to $f(Y)$ or $f(Y)^\perp$ and

$$g(Y) = f(Y) \quad \text{or} \quad g(Y) = f(Y)^\perp,$$

respectively. For each of these cases we have $g(X) \perp g(Y)$. The case when $g(X) = f(X)^\perp$ is similar. So, g is orthogonality preserving. The same holds for $g' = g^{-1}$ and g is orthogonality preserving in both directions.   □

Remark 5.13 We have to explain why the above methods cannot be exploited to study compatibility preserving transformations of $\mathcal{G}_k(H)$. Suppose that $\dim H > 3k$. If k is odd, then

$$\{X, Y\}^{cc} \cap \mathcal{G}_k(H) = \{X, Y\}$$

for any distinct compatible $X, Y \in \mathcal{G}_k(H)$. Indeed, the dimension of $(X + Y)^{\perp}$ is greater than k and the subspace

$$((X \cap Y)^{\perp} \cap X) + ((X \cap Y)^{\perp} \cap Y) \tag{5.2}$$

is even-dimensional. If k is even, then the same is true, except for the case when $\dim(X \cap Y) = k/2$. In this case, the subspace (5.2) is k-dimensional. Therefore, only for $k = 2$, we can characterize a pair of orthogonal elements of $\mathcal{G}_k(H)$ as a compatible pair $X, Y \in \mathcal{G}_k(H)$, which satisfy

$$|\{X, Y\}^{cc} \cap \mathcal{G}_k(H)| = 2.$$

If $\dim H = 6$ and $k = 2$, then

$$|\{X, Y\}^{cc} \cap \mathcal{G}_k(H)| = 3$$

for any distinct compatible $X, Y \in \mathcal{G}_k(H)$.

5.3 Proof of Theorem 5.5

Let $\{e_i\}_{i \in I}$ be an orthonormal basis of H and let $\mathcal{A}$ be the associated orthogonal apartment of $\mathcal{G}_k(H)$. Suppose that $k > 1$. As in Section 2.6, for every $i \in I$ we denote by $\mathcal{A}(+i)$ and $\mathcal{A}(-i)$ the sets consisting of all elements of $\mathcal{A}$ which contain e_i and do not contain e_i, respectively. For any distinct $i, j \in I$ we define

$$\mathcal{A}(+i, +j) = \mathcal{A}(+i) \cap \mathcal{A}(+j),$$

$$\mathcal{A}(+i, -j) = \mathcal{A}(+i) \cap \mathcal{A}(-j),$$

$$\mathcal{A}(-i, -j) = \mathcal{A}(-i) \cap \mathcal{A}(-j).$$

A subset $X \subset \mathcal{A}$ is said to be *orthogonally inexact* if there is an orthogonal apartment of $\mathcal{G}_k(H)$ distinct from $\mathcal{A}$ and containing X.

Example 5.14 For any distinct $i, j \in I$ the subset

$$\mathcal{A}(+i, +j) \cup \mathcal{A}(-i, -j) \tag{5.3}$$

is orthogonally inexact. In the basis $\{e_i\}_{i \in I}$, we replace the vectors e_i and e_j by any other pair of orthogonal unit vectors belonging to the 2-dimensional subspace spanned by e_i and e_j and distinct from scalar multiples of e_i, e_j. If $\mathcal{A}'$ is the associated orthogonal apartment of $\mathcal{G}_k(H)$, then

$$\mathcal{A} \cap \mathcal{A}' = \mathcal{A}(+i, +j) \cup \mathcal{A}(-i, -j).$$

This means that the subset (5.3) is orthogonally inexact.

Lemma 5.15 *Every orthogonally inexact subset of $\mathcal{A}$ is contained in a maximal orthogonally inexact subset and every maximal orthogonally inexact subset of $\mathcal{A}$ is of type (5.3).*

Proof We need to show that every orthogonally inexact subset $X \subset \mathcal{A}$ is contained in a subset of type (5.3). For every $i \in I$ we denote by S_i the intersection of all closed subspaces X satisfying one of the following conditions:

(1) X is an element of X containing e_i,
(2) $X^\perp$ is an element of X which does not contain e_i.

Each S_i is non-zero. If $\mathcal{A}'$ is the orthogonal apartment defined by an orthonormal basis $\{e'_i\}_{i \in I}$ and X is contained in $\mathcal{A}'$, then every subspace X satisfying (1) or (2), and consequently every S_i, is spanned by a subset of $\{e'_i\}_{i \in I}$. Therefore, if every S_i is 1-dimensional, then $\mathcal{A}$ is the unique orthogonal apartment containing X, which contradicts the assumption that X is orthogonally inexact. So, there is at least one $i \in I$ such that $\dim S_i \geq 2$. We take any $j \neq i$ such that e_j belongs to S_i and claim that

$$X \subset \mathcal{A}(+i, +j) \cup \mathcal{A}(-i, -j).$$

If $X \in X$ contains e_i, then $e_j \in S_i \subset X$ and X belongs to $\mathcal{A}(+i, +j)$. If $X \in X$ does not contain e_i, then $e_j \in S_i \subset X^\perp$, which means that e_j is not contained in X and X belongs to $\mathcal{A}(-i, -j)$.   □

We say that $C \subset \mathcal{A}$ is an *orthocomplementary* subset if $\mathcal{A} \setminus C$ is a maximal orthogonally inexact subset, i.e.

$$\mathcal{A} \setminus C = \mathcal{A}(+i, +j) \cup \mathcal{A}(-i, -j)$$

for some distinct $i, j \in I$. Then

$$C = \mathcal{A}(+i, -j) \cup \mathcal{A}(+j, -i).$$

This orthocomplementary subset will be denoted by C_{ij}. Note that $C_{ij} = C_{ji}$.

In the case when H is infinite-dimensional, there is a simple characterization of the orthogonality relation in terms of orthocomplementary subsets.

Lemma 5.16 *Suppose that H is infinite-dimensional. Then $X, Y \in \mathcal{A}$ are orthogonal if and only if the number of orthocomplementary subsets of $\mathcal{A}$ containing both X and Y is finite.*

Proof If the orthocomplementary subset C_{ij} contains both X and Y, then one of the following possibilities is realized:

(1) one of e_i, e_j belongs to $X \setminus Y$ and the other to $Y \setminus X$,
(2) one of e_i, e_j belongs to $X \cap Y$ and the other is not contained in $X + Y$.

The number of orthocomplementary subsets C_{ij} satisfying (1) is finite. If X and Y are orthogonal, then $X \cap Y = 0$ and there is no C_{ij} satisfying (2). In the case when $X \cap Y \neq 0$, the condition (2) holds for infinitely many C_{ij}. □

Suppose that H is infinite-dimensional. Let f be a bijective transformation of $\mathcal{G}_k(H)$ preserving the compatibility relation in both directions; in other words, f and f^{-1} send orthogonal apartments to orthogonal apartments. For any orthogonal k-dimensional subspaces $X, Y \subset H$ there is an orthogonal apartment $\mathcal{A} \subset \mathcal{G}_k(H)$ containing them. It is clear that f sends orthogonally inexact subsets of $\mathcal{A}$ to orthogonally inexact subsets of the orthogonal apartment $f(\mathcal{A})$. Similarly, f^{-1} transfers orthogonally inexact subsets of $f(\mathcal{A})$ to orthogonally inexact subsets of $\mathcal{A}$. This means that $\mathcal{X}$ is a maximal orthogonally inexact subset of $\mathcal{A}$ if and only if $f(\mathcal{X})$ is a maximal orthogonally inexact subset of $f(\mathcal{A})$. Therefore, a subset $C \subset \mathcal{A}$ is orthocomplementary if and only if $f(C)$ is an orthocomplementary subset of $f(\mathcal{A})$. Lemma 5.16 guarantees that $f(X)$ and $f(Y)$ are orthogonal. Similarly, we establish that f^{-1} transfers orthogonal elements of $\mathcal{G}_k(H)$ to orthogonal elements. So, f preserves the orthogonality relation in both directions and we apply Theorem 4.10.

5.4 Characterizing Lemmas

Suppose that $\dim H = n$ is finite and k is a natural number such that $1 < k \leq n - k$. Let $\mathcal{A}$ be the orthogonal apartment of $\mathcal{G}_k(H)$ defined by an orthonormal basis $\{e_i\}_{i=1}^n$.

Lemma 5.17 *Let $X, Y \in \mathcal{A}$ and $\dim(X \cap Y) = m$. Then there are precisely*

$$c(m) = (k - m)^2 + m(n - 2k + m)$$

distinct orthocomplementary subsets of $\mathcal{A}$ containing this pair.

Proof Since $\dim(X \cap Y) = m$, we have

$$\dim(X + Y) = 2k - m.$$

As in the proof of Lemma 5.16, the orthocomplementary subset C_{ij} contains both X, Y if and only if one of the following possibilities is realized:

(1) one of e_i, e_j belongs to $X \setminus Y$ and the other to $Y \setminus X$,
(2) one of e_i, e_j belongs to $X \cap Y$ and the other is not contained in $X + Y$.

There are precisely

$$(k - m)^2 \quad \text{and} \quad m(n - 2k + m)$$

distinct C_{ij} satisfying (1) and (2), respectively. □

For two distinct orthocomplementary subsets C_{ij} and $C_{i'j'}$ one of the following possibilities is realized:

- $\{i, j\} \cap \{i', j'\} = \emptyset$,
- $\{i, j\} \cap \{i', j'\} \neq \emptyset$.

In the second case, the orthocomplementary subsets are said to be *adjacent*. If C_{ij} and $C_{i'j'}$ are adjacent, then we can suppose that $i = i'$ (since $C_{ij} = C_{ji}$).

Lemma 5.18 *If the intersection of two adjacent orthocomplementary subsets of $\mathcal{A}$ and the intersection of two distinct non-adjacent orthocomplementary subsets of $\mathcal{A}$ have the same number of elements, then*

$$(n - 2k)^2 = n - 2.$$

Proof We will write

$$\mathcal{A}(+i_1, \ldots, +i_p, -j_1, \ldots, -j_q)$$

for the following intersection:

$$\mathcal{A}(+i_1) \cap \cdots \cap \mathcal{A}(+i_p) \cap \mathcal{A}(-j_1) \cap \cdots \cap \mathcal{A}(-j_q).$$

If $j \neq j'$, then

$$
\begin{aligned}
|C_{ij} \cap C_{ij'}| &= |(\mathcal{A}(+i, -j) \cup \mathcal{A}(+j, -i)) \cap (\mathcal{A}(+i, -j') \cup \mathcal{A}(+j', -i))| \\
&= |\mathcal{A}(+i, -j, -j')| + |\mathcal{A}(+j, +j', -i)| \\
&= \binom{n - 3}{k - 1} + \binom{n - 3}{k - 2} = \binom{n - 2}{k - 1}.
\end{aligned}
$$

If C_{ij} and $C_{i'j'}$ are distinct and non-adjacent, then

$$
\begin{aligned}
|C_{ij} \cap C_{i'j'}| &= |(\mathcal{A}(+i,-j) \cup \mathcal{A}(+j,-i)) \cap (\mathcal{A}(+i',-j') \cup \mathcal{A}(+j',-i'))| \\
&= |\mathcal{A}(+i,+i',-j,-j')| + |\mathcal{A}(+i,+j',-j,-i')| \\
&\quad + |\mathcal{A}(+j,+i',-i,-j')| + |\mathcal{A}(+j,+j',-i,-i')| \\
&= 4\binom{n-4}{k-2}.
\end{aligned}
$$

Suppose that the equality

$$
\binom{n-2}{k-1} = 4\binom{n-4}{k-2} \tag{5.4}
$$

holds. We have

$$
\binom{n-2}{k-1} = \binom{n-4}{k-1} + 2\binom{n-4}{k-2} + \binom{n-4}{k-3}
$$

and (5.4) implies that

$$
\binom{n-4}{k-1} + \binom{n-4}{k-3} = 2\binom{n-4}{k-2}
$$

(if $k = 2$, then $\binom{n-4}{k-3} = 0$). Simple algebraic manipulations show that

$$
(n-k-2)(n-k-1) + (k-2)(k-1) = 2(k-1)(n-k-1)
$$

and

$$
(n-2k-1)(n-k-1) = (k-1)(n-2k+1).
$$

We rewrite the latter equality as

$$
(n-2k-1)(n-2k) = 2(k-1)
$$

and establish that $(n-2k)^2 = n-2$. $\qquad\square$

For any distinct $X, Y \in \mathcal{A}$ we denote by $\mathfrak{C}(X, Y)$ the family of all orthocomplementary subsets of $\mathcal{A}$ containing both X and Y.

Lemma 5.19 *Subspaces $X, Y \in \mathcal{A}$ are orthogonal if and only if for every orthocomplementary subset $C \in \mathfrak{C}(X, Y)$ there are precisely $2(k-1)$ elements of $\mathfrak{C}(X, Y)$ adjacent to C.*

Proof Consider any $C_{ij} \in \mathfrak{C}(X, Y)$ such that $e_i \in X \setminus Y$ and $e_j \in Y \setminus X$. An orthocomplementary subset from $\mathfrak{C}(X, Y)$ is adjacent to C_{ij} if and only if it coincides with $C_{i'j}$ or $C_{ij'}$, where $e_{i'} \in X \setminus Y$ and $e_{j'} \in Y \setminus X$. There are precisely $2(m-1)$ such orthocomplementary subsets, where $m = k - \dim(X \cap Y)$.

If X and Y are orthogonal, then $m = k$ and for every $C_{ij} \in \mathfrak{C}(X, Y)$ one of the vectors e_i, e_j belongs to $X \setminus Y$ and the other to $Y \setminus X$. Therefore, for each $C \in \mathfrak{C}(X, Y)$ there are precisely $2(k-1)$ elements of $\mathfrak{C}(X, Y)$ adjacent to C.

If X and Y are non-orthogonal, then $m < k$ and $\mathfrak{C}(X, Y)$ contains an ortho-complementary subset adjacent to precisely $2(m-1)$ elements of $\mathfrak{C}(X, Y)$. $\square$

5.5 Proofs of Theorems 5.6 and 5.7

As in the previous section, we suppose that $\dim H = n$ is finite. Let f be a compatibility preserving injective transformation of $\mathcal{G}_k(H)$. Then f sends orthogonal apartments to orthogonal apartments. It was noted in Section 5.1 that we can restrict ourselves to the case when $1 < k < n - k$.

For any two compatible elements of $\mathcal{G}_k(H)$ there is an orthogonal apartment containing them. Therefore, it is sufficient to prove that the restriction of f to every orthogonal apartment is orthogonality and adjacency preserving. This implies that f is an orthogonality and ortho-adjacency preserving injection and Theorem 4.26 gives the claim.

Let $\mathcal{A}$ be an orthogonal apartment of $\mathcal{G}_k(H)$. Then $f(\mathcal{A})$ is an orthogonal apartment of $\mathcal{G}_k(H)$ and f sends orthogonally inexact subsets of $\mathcal{A}$ to orthogonally inexact subsets of $f(\mathcal{A})$. Observe that $\mathcal{A}$ and $f(\mathcal{A})$ have the same finite number of orthogonally inexact subsets. This means that X is an orthogonally inexact subset of $\mathcal{A}$ if and only if $f(X)$ is an orthogonally inexact subset of $f(\mathcal{A})$. Then an orthogonally inexact subset $X \subset \mathcal{A}$ is maximal if and only if the same holds for $f(X)$. Consequently, a subset $C \subset \mathcal{A}$ is orthocomplementary if and only if $f(C)$ is an orthocomplementary subset of $f(\mathcal{A})$. In particular,

$$f(\mathfrak{C}(X, Y)) = \mathfrak{C}(f(X), f(Y))$$

for all $X, Y \in \mathcal{A}$.

Let us consider the quadratic function

$$c(x) = (k-x)^2 + x(n - 2k + x) = 2x^2 - (4k - n)x + k^2.$$

It takes the minimum value at $x = (4k-n)/4$. By Lemma 5.17, for any $X, Y \in \mathcal{A}$ we have

$$|\mathfrak{C}(X, Y)| = c(m),$$

where $m = \dim(X \cap Y)$.

If $n \geq 4k$, then $(4k - n)/4 \leq 0$ and

$$c(0) < c(1) < \cdots < c(k-1),$$

which implies that the restriction of f to $\mathcal{A}$ is orthogonality and adjacency pre-serving, i.e. f is orthogonality and ortho-adjacency preserving. So, Theorem 5.7 is proved for $n \geq 4k$.

Consider the case when $n < 4k$. Since the function $c(x)$ takes the minimal value at $x = (4k - n)/4 > 0$, we have

$$c\left(\frac{4k - n}{4} - x\right) = c\left(\frac{4k - n}{4} + x\right) \text{ and } c(0) = c\left(\frac{4k - n}{2}\right).$$

If n is odd, then $(4k - n)/2$ is not natural and f is orthogonality preserving. In the case when n is even and $(n - 2k)^2 \neq n - 2$, Lemma 5.18 implies that two orthocomplementary subsets of $\mathcal{A}$ are adjacent if and only if their images are adjacent orthocomplementary subsets of $f(\mathcal{A})$. By Lemma 5.19, this guar-antees that for any $X, Y \in \mathcal{A}$ the images $f(X)$ and $f(Y)$ are orthogonal if and only if X and Y are orthogonal. Therefore, f is orthogonality preserving if the possibility described in Theorem 5.7 is not realized (we cannot state that f is orthogonality preserving in both directions and apply Theorem 4.29, since for an orthogonal apartment $\mathcal{A} \subset \mathcal{G}_k(H)$ the preimage $f^{-1}(\mathcal{A})$ is not necessarily an orthogonal apartment). The following lemma completes the proof of Theorem 5.7 for the case when $n < 4k$.

Lemma 5.20 *If $n < 4k$, then f is ortho-adjacency preserving under the as-sumption that the possibility $n - 6, k = 2$ is not realized.*

Proof We have

$$c(k - 1) > c(m)$$

for all $m \in \{0, 1, \ldots, k - 2\}$ if

$$k - 1 > \frac{4k - n}{2}$$

or, equivalently, if $k < (n - 2)/2$. In this case, two elements of $\mathcal{A}$ are adjacent if and only if their images are adjacent elements of $f(\mathcal{A})$. By our assumption, $n = 2k + l$ for a certain natural $l > 0$ and the inequality $k < (n - 2)/2$ does not hold only in the case when $l \in \{1, 2\}$.

If $n = 2k + 2$, then $k - 1$ is equal to $(4k - n)/2$ and we have

$$c(k - 1) = c(0) > c(m)$$

for every $m \in \{1, \ldots, k - 2\}$. If $(n - 2k)^2 \neq n - 2$, then $X, Y \in \mathcal{A}$ are orthogonal if and only if $f(X)$ and $f(Y)$ are orthogonal (by Lemmas 5.18 and 5.19), which implies that two elements of $\mathcal{A}$ are adjacent if and only if their images are adjacent elements of $f(\mathcal{A})$. Since $n = 2k + 2$, the equality $(n - 2k)^2 = n - 2$ holds only in the case when $n = 6$ and $k = 2$.

If $n = 2k + 1$, then we have

$$c(k - 1) \neq c(m)$$

for every $m \in \{0, 1, \ldots, k - 2\}$. Indeed, if $c(k - 1) = c(x)$ and $x \neq k - 1$, then an easy calculation shows that $x = 1/2$. This means that $X, Y \in \mathcal{A}$ are adjacent if and only if $f(X)$ and $f(Y)$ are adjacent. □

Now, we prove Theorem 5.6. We will need the following remarkable result.

Theorem 5.21 (Westwick [66]) *Let V be a vector space of finite dimension (over a division ring). Then every adjacency preserving bijective transformation of $\mathcal{G}_k(V)$ is adjacency preserving in both directions, i.e. it is an automorphism of the Grassmann graph $\Gamma_k(V)$.*

The case when $k = 1, \dim V - 1$ is trivial. See [45] for more information.

Proof of Theorem 5.6 Suppose that f is bijective and the case $n = 6, k = 2$ is not realized. It was established above that f is ortho-adjacency preserving (but we cannot state that f is ortho-adjacency preserving in both directions). Every ortho-adjacency preserving injective transformation of $\mathcal{G}_k(H)$ is adjacency preserving if $\dim H > 2k > 2$ (by Lemma 4.37). So, f is an adjacency preserving bijection. Theorem 5.21 says that f is adjacency preserving in both directions, i.e. f is induced by a semilinear automorphism of H. Since f is ortho-adjacency preserving, this semilinear isomorphism sends orthogonal vectors to orthogonal vectors. By Proposition 4.2, it is a scalar multiple of a unitary or anti-unitary operator. □

Remark 5.22 Suppose that $n = 6$ and $k = 2$. As above, we assume that $\mathcal{A}$ is an orthogonal apartment of $\mathcal{G}_k(H)$. Any two distinct $X, Y \in \mathcal{A}$ are contained in precisely four distinct orthocomplementary subsets of $\mathcal{A}$. The intersection of any two distinct orthocomplementary subsets consists of four elements. Thus the dimension of $X \cap Y$ cannot be determined in terms of orthocomplementary subsets. By Remark 5.13, we have

$$|\{X, Y\}^{cc} \cap \mathcal{G}_k(H)| = 3$$

for any $X, Y \in \mathcal{A}$ and the methods from Section 5.2 also cannot be exploited. It is an open problem to extend Theorem 5.6 to the case when $n = 6$ and k is equal to 2 or 4.

5.6 Proof of Theorem 5.9

Suppose that $\dim H = 2k \geq 8$ and f is a bijective transformation of $\mathcal{G}_k(H)$ preserving the compatibility relation in both directions. Then f and f^{-1} send orthogonal apartments to orthogonal apartments.

In this case, we have

$$c(m) = (k - m)^2 + m^2.$$

Then $c(m) = c(m')$ implies that $m' = m$ or $m' = k - m$. Using arguments from the previous section, we prove the following.

Lemma 5.23 *If X and Y are compatible elements of $\mathcal{G}_k(H)$ and $X \cap Y$ is m-dimensional, then the dimension of $f(X) \cap f(Y)$ is equal to m or $k - m$.*

We put $m = 0$ in the previous lemma and get the following.

Lemma 5.24 $f(X^{\perp}) = f(X)^{\perp}$ *for every $X \in \mathcal{G}_k(H)$.*

Denote by $\mathcal{G}'$ the set of all two-element subsets $\{X, X^{\perp}\} \subset \mathcal{G}_k(H)$. Consider the graph Γ' whose vertex set is $\mathcal{G}'$ and two pairs $\{X, X^{\perp}\}$ and $\{Y, Y^{\perp}\}$ are adjacent vertices in this graph if X is adjacent to Y or $Y^{\perp}$. The latter condition implies that $X^{\perp}$ is adjacent to $Y^{\perp}$ or Y, respectively. Since $k > 2$, only one of the subspaces $Y, Y^{\perp}$ is adjacent to X (if $n = 4$, then there are 2-dimensional subspaces $X, Y \subset H$ such that X is adjacent to both Y and $Y^{\perp}$). In what follows, two elements of $\mathcal{G}'$ will be called *adjacent* if they are adjacent vertices of the graph Γ'.

For every $(k - 1)$-dimensional subspace $S \subset H$ we denote by $C(S)$ the set of all $\{X, X^{\perp}\} \in \mathcal{G}'$ such that X or $X^{\perp}$ contains S. Observe that $\{X, X^{\perp}\}$ belongs to $C(S)$ if and only if the $(k + 1)$-dimensional subspace $S^{\perp}$ contains $X^{\perp}$ or X. It is clear that $C(S)$ is a clique of the graph Γ'.

Lemma 5.25 *Every maximal clique of Γ' is $C(S)$ with $S \in \mathcal{G}_{k-1}(H)$.*

Proof Suppose that C is a clique of Γ' containing more than one element. Let $\{X, X^{\perp}\} \in C$. It was noted above that for every $\{Y, Y^{\perp}\} \in C$ only one of the subspaces is adjacent to X. We denote by $\mathcal{X}$ the set containing X and all $Y \in \mathcal{G}_k(H)$ such that $\{Y, Y^{\perp}\} \in C$ and Y is adjacent to X.

Let Y and Y' be distinct elements of $\mathcal{X} \setminus \{X\}$. If Y and Y' are not adjacent, then

$$\dim(Y \cap Y') = k - 2$$

(since Y and Y' both are adjacent to X) and

$$\dim(Y \cap Y'^{\perp}) \leq 2 < k - 1$$

(since $k \geq 4$). Therefore, Y is not adjacent to any of Y', $Y'^{\perp}$, which contradicts the fact that $\{Y, Y^{\perp}\}$ and $\{Y', Y'^{\perp}\}$ are adjacent elements of $\mathcal{G}'$. So, X is a clique of the Grassmann graph $\Gamma_k(H)$.

Let $\mathcal{Y}$ be a maximal clique of $\Gamma_k(H)$ containing X. If $\mathcal{Y}$ is the star corresponding to a $(k-1)$-dimensional subspace S, then C is contained in $C(S)$. If $\mathcal{Y}$ is the top defined by a $(k+1)$-dimensional subspace N, then the orthogonal complement $N^{\perp}$ is $(k-1)$-dimensional and C is a subset of $C(N^{\perp})$. $\square$

Lemma 5.26 *If X, Y, Z are mutually ortho-adjacent elements of $\mathcal{G}_k(H)$, then there is a unique maximal clique of Γ' containing the corresponding elements of $\mathcal{G}'$.*

Proof For the subspaces

$$S = X \cap Y \cap Z \ \text{ and } \ N = X + Y + Z$$

one of the following possibilities is realized:

(1) $\dim S = k - 1$ and $\dim N = k + 2$,
(2) $\dim S = k - 2$ and $\dim N = k + 1$.

The required maximal clique is $C(S)$ or $C(N^{\perp})$, respectively. $\square$

Remark 5.27 If X, Y are adjacent elements of $\mathcal{G}_k(H)$, then the corresponding elements of $\mathcal{G}'$ are contained in the maximal cliques

$$C(X \cap Y) \ \text{ and } \ C((X + Y)^{\perp}).$$

Also, for every $Z \in \mathcal{G}_k(H)$ belonging to the line joining X and Y the corresponding element of $\mathcal{G}'$ is contained in each of these maximal cliques.

Lemma 5.24 shows that f induces a bijective transformation of $\mathcal{G}'$. We denote this transformation by f'.

Lemma 5.28 *The transformation f' is an automorphism of the graph Γ'.*

Proof Let $\{X, X^{\perp}\}$ and $\{Y, Y^{\perp}\}$ be adjacent elements of $\mathcal{G}'$. We need to show that f' transfers them to adjacent elements of $\mathcal{G}'$. It is sufficient to restrict ourselves to the case when X and Y are adjacent.

Suppose that X, Y are ortho-adjacent. Then there is an orthogonal apartment $\mathcal{A} \subset \mathcal{G}_k(H)$ containing them. Note that $X^{\perp}$ and $Y^{\perp}$ also belong to this apartment. By Lemma 5.23, $f(X)$ and $f(Y)$ are ortho-adjacent or

$$\dim(f(X) \cap f(Y)) = 1.$$

Since the orthogonal apartment $f(\mathcal{A})$ contains $f(X), f(Y)$ and their orthogonal complements, the latter equality implies that $f(X)$ is ortho-adjacent to

$f(Y)^\perp = f(Y^\perp)$. Therefore, f' transfers $\{X, X^\perp\}$ and $\{Y, Y^\perp\}$ to adjacent elements of $\mathcal{G}'$; moreover, the images of $\{X, X^\perp\}$ and $\{Y, Y^\perp\}$ are defined by ortho-adjacent elements of $\mathcal{G}_k(H)$.

Consider the case when X and Y are non-compatible. Since the dimension of $(X + Y)^\perp$ is equal to $k - 1 \geq 3$, this subspace contains three mutually orthogonal 1-dimensional subspaces P, Q, T. We set

$$X' = P + (X \cap Y), \quad Y' = Q + (X \cap Y), \quad Z' = T + (X \cap Y).$$

Observe that X, X', Y', Z' are mutually ortho-adjacent and the same holds for Y, X', Y', Z'. Then the elements of $\mathcal{G}'$ corresponding to

$$f(X), f(X'), f(Y'), f(Z')$$

are mutually adjacent, i.e. there is a maximal clique $C(M)$, $M \in \mathcal{G}_{k-1}(H)$ containing them. Similarly, there is a maximal clique $C(N)$, $N \in \mathcal{G}_{k-1}(H)$ containing the elements of $\mathcal{G}'$ corresponding to

$$f(Y), f(X'), f(Y'), f(Z').$$

Then $C(M) \cap C(N)$ contains the elements of $\mathcal{G}'$ defined by $f(X'), f(Y'), f(Z')$. Since these elements of $\mathcal{G}'$ can be obtained from mutually ortho-adjacent elements of $\mathcal{G}_k(H)$, Lemma 5.26 implies that $M = N$. So, the elements of $\mathcal{G}'$ corresponding to $f(X)$ and $f(Y)$ belong to a certain maximal clique of Γ', i.e. they are adjacent.

Since f and f^{-1} send orthogonal apartments to orthogonal apartments, the same arguments show that f'^{-1} is adjacency preserving. □

Therefore, f' and f'^{-1} transfer maximal cliques of Γ' to maximal cliques, i.e. there is a bijective transformation g of $\mathcal{G}_{k-1}(H)$ such that

$$f'(C(S)) = C(g(S))$$

for every $S \in \mathcal{G}_{k-1}(H)$.

Lemma 5.29 *The transformation g is compatibility preserving.*

Proof It is sufficient to show that g sends orthogonal apartments to orthogonal apartments. Let B be an orthonormal basis of H and let $\mathcal{A}$ be the associated orthogonal apartment of $\mathcal{G}_k(H)$. Suppose that the orthogonal apartment $f(\mathcal{A})$ is defined by an orthonormal basis B'. Then g transfers the orthogonal apartment of $\mathcal{G}_{k-1}(H)$ defined by B to the orthogonal apartment of $\mathcal{G}_{k-1}(H)$ associated to the basis B'. □

Since $n > 6$, Theorem 5.6 implies that g can be uniquely extended to an automorphism of $\mathcal{L}(H)$ induced by a unitary or anti-unitary operator. This extension also will be denoted by g.

Let $X \in \mathcal{G}_k(H)$. We take any $S, N \in \mathcal{G}_{k-1}(H)$ such that $X = S + N$. Then

$$C(S) \cap C(N) = \{X, X^{\perp}\}$$

(since $k > 2$, for $k = 2$ this fails) and

$$f'(C(S) \cap C(N)) = C(g(S)) \cap C(g(N)) = \{X', X'^{\perp}\},$$

where

$$X' = g(S) + g(N) = g(X)$$

(since g is an automorphism of $\mathcal{L}(H)$). This means that f' transfers $\{X, X^{\perp}\}$ to $\{g(X), g(X)^{\perp}\}$. Therefore, $f(X)$ coincides with $g(X)$ or $g(X)^{\perp}$.

Remark 5.30   Several times we use the assumptions that $k \geq 4$ and f is compatibility preserving in both directions. It is an open problem to extend Theorem 5.9 to the case when $k = 2, 3$. Is it possible to omit the assumption that f^{-1} is compatibility preserving?

5.7 A Metric Connected to the Compatibility Relation

Following [39], we consider the *separation* metric on the lattice $\mathcal{L}(H)$. This metric on orthomodular lattices was first studied in [56].

For any two closed subspaces $X, Y \subset H$ we define

$$s(X, Y) = \|P_X + P_Y - 2P_{X \cap Y}\|.$$

It is clear that

$$s(X, Y) = \|P_{X'} + P_{Y'}\|,$$

where

$$X' = (X \cap Y)^{\perp} \cap X \quad \text{and} \quad Y' = (X \cap Y)^{\perp} \cap Y.$$

Therefore, $s(X, Y) = 0$ if and only if $X \cap Y$ coincides with both X and Y, i.e. only in the case when $X = Y$. For distinct X, Y we have

$$1 \leq s(X, Y) \leq 2,$$

which implies immediately the fulfilment of the triangle inequality. So, g is a metric. By [18, 64],

$$s(X, Y) = \|P_{X'} + P_{Y'}\| = 1 + \|P_{X'} P_{Y'}\|,$$

which means that $s(X, Y) = 1$ if and only if X' and Y' are orthogonal (possibly, one of X', Y' is zero), i.e. X and Y are compatible.

Suppose that $s(X, Y) > 1$, i.e. X and Y are non-compatible. If X', Y' are finite-dimensional, then $s(X, Y) = 1 + \cos^2 \theta$, where θ is the smallest principal angle between X' and Y'. It can take any value from $(0, \pi/2)$ and we obtain that $s(X, Y) < 2$.

If H is infinite-dimensional, then there are pairs of closed infinite-dimensional subspaces $X, Y \subset H$ such that $s(X, Y) = 2$.

Example 5.31 Without loss of generality we can suppose that $H = l_2(\mathbb{C})$. Let X and Y be the subspaces formed by all sequences from $l_2(\mathbb{C})$ of the forms

$$(a_1, 0, a_3, 0, a_5, \dots) \quad \text{and} \quad (b_1, b_1/2, b_2, b_2/4, b_3, b_3/6, \dots),$$

respectively. Then $X \cap Y = 0$ and $s(X, Y) = \|P_X + P_Y\|$. Denote by x_k the element of $l_2(\mathbb{C})$ whose $(2k - 1)$-coordinate is 1, the $(2k)$-coordinate is $1/(2k)$ and all other coordinates are zero. Then

$$\frac{\|P_X(x_k) + P_Y(x_k)\|^2}{\|x_k\|^2} = \frac{4 + 1/(4k^2)}{1 + 1/(4k^2)}$$

converges to 4 if $k \to \infty$ and we have $s(X, Y) = 2$.

Every automorphism of $\mathcal{L}(H)$ induced by a unitary or anti-unitary operator preserves the separation metric. By [39], we have $s(X^\perp, Y^\perp) = s(X, Y)$ for any $X, Y \in \mathcal{L}(H)$. In the case when H is finite-dimensional, the latter equality follows from the fact that the smallest non-zero principal angle between X and Y is equal to the smallest non-zero principal angle between $X^\perp$ and $Y^\perp$.

As a consequence of Theorem 5.2, we get the following.

Corollary 5.32 *Let f be a bijective transformation of $\mathcal{L}(H)$ preserving the separation metric. Then the restriction of f to $\mathcal{L}(H) \setminus \{0, H\}$ coincides with the restriction of an automorphism or anti-automorphism of $\mathcal{L}(H)$ induced by a unitary or anti-unitary operator.*

Remark 5.33 Note that 0 and H can be characterized as elements of $\mathcal{L}(H)$ which are at distance one from all elements of $\mathcal{L}(H)$ except themselves. Then for every bijective transformation f of $\mathcal{L}(H)$ preserving the separation metric we have $f(\{0, H\}) = \{0, H\}$ and each of the possibilities $f(0) = 0$, $f(H) = H$ or $f(0) = H$, $f(H) = 0$ is realized.

Proof The transformation f is compatibility preserving in both directions and Theorem 5.2 implies the existence of an automorphism g of $\mathcal{L}(H)$ induced by a unitary or anti-unitary operator and such that for every $X \in \mathcal{L}(H)$ we have either

$$f(X) = g(X) \quad \text{or} \quad f(X) = g(X)^{\perp}.$$

The transformation $h = g^{-1}f$ preserves the separation metric and every $h(X)$ coincides with X or $X^{\perp}$.

Suppose that $h(P) = P$ for a certain 1-dimensional subspace $P \subset H$ and consider any $X \in \mathcal{L}(H) \setminus \{0, H\}$. In the case when X is non-compatible to P, we have $0 < \angle(P, X) < \pi/2$ and

$$s(P, X) = 1 + \cos^2 \angle(P, X).$$

Since h preserves the separation metric, we obtain that

$$\angle(P, X) = \angle(P, h(X)).$$

If $\angle(P, X) \neq \pi/4$, then $\angle(P, X) \neq \angle(P, X^{\perp})$, which means that $h(X) = X$.

In the case when $\angle(P, X) = \pi/4$ or P and X are compatible, we choose a 1-dimensional subspace $Q \subset H$ such that each of the angles $\angle(P, Q)$ and $\angle(Q, X)$ is distinct from $0, \pi/4, \pi/2$. Then $h(Q) = Q$, which implies that $h(X) = X$.

If h sends every 1-dimensional subspace to the orthogonal complement, then we apply the above arguments to the composition of h and the orthocomplementation and obtain that $h(X) = X^{\perp}$ for every $X \in \mathcal{L}(H) \setminus \{0, H\}$. □

6

Applications

In this final chapter, we present some applications of Wigner's theorem and its generalizations obtained in Chapter 4.

The first is the classic Kadison's theorem [30] concerning automorphisms of the convex set of all bounded positive operators of trace one. All bijective transformations preserving the convex structure of this set are induced by unitary and anti-unitary operators.

In the second section, we consider the real vector space formed by all self-adjoint operators of finite rank on a complex Hilbert space (of dimension not less than three). It is not difficult to prove that linear automorphisms of this vector space preserving projections are induced by unitary and anti-unitary operators. Using some arguments from the proofs of Theorems 4.25 and 4.26, we describe linear transformations sending projections of fixed rank k to projections of rank k, as well as linear transformations which map projections of a fixed rank to projections of other fixed rank.

6.1 Automorphisms of the Convex Set of Quantum States

Let H be a complex Hilbert space. In Section 1.3, we consider the convex set $\mathcal{J}(H)$ formed by all positive trace class operators $A \in \mathcal{B}(H)$ satisfying $\operatorname{tr}(A) = 1$. The set of extreme points for this convex set is $\mathcal{P}_1(H)$. By Gleason's theorem, if H is separable and $\dim H \geq 3$, then $\mathcal{J}(H)$ can be identified with the convex set of all states of the orthomodular lattice $\mathcal{L}(H)$ and rank-one projections correspond to pure states. Using Proposition 4.8 (Uhlhorn's version of Wigner's theorem), we describe all bijective transformations of $\mathcal{J}(H)$ preserving the convex structure.

Theorem 6.1 (Kadison [30]) *Suppose that* $\dim H \geq 3$ *and F is a bijective* *transformation of $\mathcal{J}(H)$ satisfying*

$$F(tA + (1 - t)B) = tF(A) + (1 - t)F(B)$$

for all $A, B \in \mathcal{J}(H)$ and $t \in [0, 1]$. Then there is a unitary or anti-unitary operator U (unique up to a scalar multiple of modulo one) such that

$$F(A) = UAU^*$$

for every $A \in \mathcal{J}(H)$.

The proof of Theorem 6.1 is given after Lemma 6.2, and taken from [63, Section IV.3] (with some modifications). We do not assume that H is separable. Recall that for a trace class operator $A \in \mathcal{B}(H)$ we have

$$\text{tr}(A) = \sum_{i \in I} \langle A(e_i), e_i \rangle$$

for any orthonormal basis $\{e_i\}_{i \in I}$ of H. As in [63], we denote by $\mathcal{W}_+$ the set of all bounded positive trace class operators on H. For $A, B \in \mathcal{W}_+$ we write $A \geq B$ if $A - B$ is positive.

Every bijective transformation F of $\mathcal{J}(H)$ satisfying the condition of Theorem 6.1 can be uniquely extended to a bijective transformation of $\mathcal{W}_+$ (we denote this extension by the same symbol F) such that the following assertions are fulfilled:

(1) $\text{tr}(F(A)) = \text{tr}(A)$ for every $A \in \mathcal{W}_+$;
(2) $F(aA + bB) = aF(A) + bF(B)$ for all $A, B \in \mathcal{W}_+$ and all positive real numbers a, b;
(3) for $A, B \in \mathcal{W}_+$ we have $A \geq B$ if and only if $F(A) \geq F(B)$.

For every non-zero $A \in \mathcal{W}_+$ there is a unique operator $A' \in \mathcal{J}(H)$ such that $A = \text{tr}(A)A'$ and we define

$$F(A) = \text{tr}(A)F(A').$$

The property (1) is obvious. To prove (2) it is sufficient to consider $aA + bB$ as

$$[a \cdot \text{tr}(A) + b \cdot \text{tr}(B)] \frac{a \cdot \text{tr}(A)A' + b \cdot \text{tr}(B)B'}{a \cdot \text{tr}(A) + b \cdot \text{tr}(B)},$$

where A' and B' belong to $\mathcal{J}(H)$. If $A, B \in \mathcal{W}_+$ and $A \geq B$, then $A - B \in \mathcal{W}_+$ and we have

$$F(A) = F(B) + F(A - B) \geq F(B);$$

applying the same arguments to F^{-1}, we establish (3).

The proof of Theorem 6.1 is based on the following characterization of the orthogonality relation.

Lemma 6.2 *Two distinct 1-dimensional subspaces $X, Y \subset H$ are orthogonal if and only if for any $A \in \mathcal{W}_+$ the relations $A \geq P_X$ and $A \geq P_Y$ imply that $\mathrm{tr}(A) \geq 2$.*

Proof Suppose that $X, Y \in \mathcal{G}_1(H)$ are orthogonal. Consider an orthonormal basis $\{e_1, e_2\} \cup \{e_i\}_{i \in I}$ of H such that $e_1 \in X$ and $e_2 \in Y$. If $A \geq P_X$ and $A \geq P_Y$ for a certain $A \in \mathcal{W}_+$, then

$$\mathrm{tr}(A) = \langle A(e_1), e_1 \rangle + \langle A(e_2), e_2 \rangle + \sum_{i \in I} \langle A(e_i), e_i \rangle$$

$$\geq \langle A(e_1), e_1 \rangle + \langle A(e_2), e_2 \rangle \geq \langle P_X(e_1), e_1 \rangle + \langle P_Y(e_2), e_2 \rangle = 2.$$

In the case when $X, Y \in \mathcal{G}_1(H)$ are non-orthogonal, we need to establish the existence of $A \in \mathcal{W}_+$ satisfying $A \geq P_X, A \geq P_Y$ and such that $\mathrm{tr}(A) < 2$. Since X and Y are distinct, their sum is a 2-dimensional subspace and we can restrict ourselves to the case when this subspace coincides with H.

So, we assume that X and Y are non-orthogonal 1-dimensional subspaces and $H = X + Y$. We can choose unit vectors $x \in X$, $y \in Y$ and a unit vector z orthogonal to x such that $y = ax + bz$, where a and b are positive real numbers satisfying $a^2 + b^2 = 1$; each of these numbers is non-zero and less than one. With respect to the orthogonal basis $\{x, z\}$ the operator $P_Y - P_X$ has the following matrix:

$$\begin{pmatrix} a^2 & ab \\ ab & b^2 \end{pmatrix} - \begin{pmatrix} 1 & 0 \\ 0 & 0 \end{pmatrix} = \begin{pmatrix} a^2 - 1 & ab \\ ab & b^2 \end{pmatrix}.$$

Taking into account the equality $a^2 + b^2 = 1$, we establish that the eigenvalues of this matrix are b and $-b$. Therefore, there exists a unitary operator U such that the matrix of the operator $U(P_Y - P_X)U^*$ (with respect to the basis $\{x, z\}$) is

$$\begin{pmatrix} b & 0 \\ 0 & -b \end{pmatrix}.$$

Consider the operator B such that the matrix of the operator UBU^* (with respect to the basis $\{x, z\}$) is

$$\begin{pmatrix} b & 0 \\ 0 & 0 \end{pmatrix}.$$

This operator is positive and $\mathrm{tr}(B) = b < 1$. The operator

$$UBU^* - U(P_Y - P_X)U^*$$

is positive and we have $B \geq P_Y - P_X$. The operator $A = B + P_X$ is as required. Indeed, it is clear that $A \geq P_X, A \geq P_Y$ and $\mathrm{tr}(A) = 1 + b < 2$. $\square$

Proof of Theorem 6.1 Since $\mathcal{P}_1(H)$ is the set of extreme points of $\mathcal{J}(H)$, we have

$$F(\mathcal{P}_1(H)) = \mathcal{P}_1(H)$$

and F induces a bijective transformation f of $\mathcal{G}_1(H)$. By Lemma 6.2, this transformation is orthogonality preserving in both directions. In the case when $\dim H \geq 3$, Proposition 4.8 implies that f is induced by a unitary or anti-unitary operator U. Then $F(A) = UAU^*$ for every rank-one projection A. We need to show that the same holds for all $A \in \mathcal{W}^+$.

Every $A \in \mathcal{W}^+$ can be presented as a series $\sum_{i \in I} t_i P_i$, where I is a countable set, $\{P_i\}_{i \in I}$ are rank-one projections with mutually orthogonal images and $\{t_i\}_{i \in I}$ are positive real numbers satisfying $\sum_{i \in I} t_i = \operatorname{tr}(A)$. If I is finite, then the required statement follows from the property (2). Suppose that I is infinite. Since $\{F(P_i)\}_{i \in I}$ are rank-one projections with mutually orthogonal images, the series $\sum_{i \in I} t_i F(P_i)$ converges to a certain $B \in \mathcal{W}^+$ whose trace is equal to $\operatorname{tr}(A) = \operatorname{tr}(F(A))$. For every finite subset $J \subset I$ we have $A \geq \sum_{i \in J} t_i P_i$ and, consequently,

$$F(A) \geq F\left(\sum_{i \in J} t_i P_i \right) = \sum_{i \in J} t_i F(P_i),$$

which implies that $F(A) \geq B$. Then $F(A) = B$, since these operators have the same trace.   □

6.2 Linear Transformations which Preserve Projections

As in the previous section, we suppose that H is a complex Hilbert space. Recall that the real vector space $\mathcal{F}_s(H)$ of all finite-rank self-adjoint operators on H is spanned by the set of all rank-k projections $\mathcal{P}_k(H)$ for every finite $k < \dim H$.

Consider a few examples of linear transformations of $\mathcal{F}_s(H)$ which send projections to projections.

Example 6.3 Let U be a linear or conjugate-linear isometry of H to itself. Then U^*U is identity and U^* is surjective. The transformation L_U of $\mathcal{F}_s(H)$ defined as

$$L_U(A) = UAU^* \quad \text{for all} \ \ A \in \mathcal{F}_s(H)$$

is a linear injection which sends the projection on a closed subspace X to the projection on $U(X)$, i.e. it transfers every $\mathcal{P}_k(H)$ to a subset of $\mathcal{P}_k(H)$. This is a linear automorphism of $\mathcal{F}_s(H)$ only in the case when U is a unitary or anti-unitary operator.

Example 6.4 Suppose that $\dim H = n$ is finite. We fix k and consider the transformation $L_k^\perp$ of $\mathcal{F}_s(H)$ defined as follows:

$$L_k^\perp(A) = k^{-1}\mathrm{tr}(A)\mathrm{Id}_H - A \quad \text{for all } A \in \mathcal{F}_s(H).$$

This is a linear automorphism of $\mathcal{F}_s(H)$ which sends the projection on a k-dimensional subspace X to the projection on the orthogonal complement $X^\perp$, i.e. it transfers $\mathcal{P}_k(H)$ to $\mathcal{P}_{n-k}(H)$. If Y is an m-dimensional subspace of H and $m \neq k$, then $L_k^\perp(P_Y)$ is not a projection. In the case when $n = 2k$, the linear automorphism $L_k^\perp$ sends $\mathcal{P}_k(H)$ to itself.

Example 6.5 We fix a rank-k projection P and consider the transformation of $\mathcal{F}_s(H)$ defined as

$$A \to k^{-1}\mathrm{tr}(A)P$$

for every $A \in \mathcal{F}_s(H)$. This is a linear transformation whose image is the 1-dimensional subspace containing P. It maps every rank-k projection to P.

Theorem 6.6 *If* $\dim H \geq 3$ *and L is a linear automorphism of $\mathcal{F}_s(H)$ preserving the set of projections in both directions, i.e. $P \in \mathcal{F}_s(H)$ is a projection if and only if $L(P)$ is a projection, then $L = L_U$ for a certain unitary or anti-unitary operator U.*

Proof First of all we observe that closed subspaces $X, Y \subset H$ are orthogonal if and only if $P_X + P_Y$ is a projection. Indeed, if X and Y are orthogonal, then $P_X + P_Y = P_{X+Y}$. Conversely, suppose that $P_X + P_Y = P_Z$ for a certain closed subspace $Z \subset H$. Lemma 4.42 states that $Z = X + Y$. Then $P_Z = P_X + P_{Y'}$, where Y' is the orthogonal complement of X in Z, and we have $Y = Y'$.

A projection belongs to $\mathcal{F}_s(H)$ if and only if its image is finite-dimensional. Consider the lattice $\mathcal{L}_{\mathrm{fin}}(H)$ formed by all finite-dimensional subspaces of H and the bijective transformation f of $\mathcal{L}_{\mathrm{fin}}(H)$ induced by L, i.e.

$$L(P_X) = P_{f(X)}$$

for every $X \in \mathcal{L}_{\mathrm{fin}}(H)$. If X and Y are orthogonal elements of $\mathcal{L}_{\mathrm{fin}}(H)$, then $P_{X+Y} = P_X + P_Y$ and

$$P_{f(X+Y)} = L(P_{X+Y}) = L(P_X) + L(P_Y) = P_{f(X)} + P_{f(Y)},$$

which means that $f(X)$ and $f(Y)$ are orthogonal. Applying the same arguments to L^{-1} and f^{-1}, we establish that f is orthogonality preserving in both directions.

For every $X \in \mathcal{L}_{\text{fin}}(H)$ we denote by $\text{ort}_{\text{fin}}(X)$ the set of all elements of $\mathcal{L}_{\text{fin}}(H)$ orthogonal to X. Then

$$X \subset Y \iff \text{ort}_{\text{fin}}(Y) \subset \text{ort}_{\text{fin}}(X)$$

for $X, Y \in \mathcal{L}_{\text{fin}}(H)$. As in the proof of Theorem 4.4, we show that f is an automorphism of the lattice $\mathcal{L}_{\text{fin}}(H)$. Then the restriction of f to $\mathcal{G}_1(H)$ is an automorphism of the projective space Π_H, which means that f is induced by a semilinear automorphism of H. Since f is orthogonality preserving, this semi-linear automorphism is a scalar multiple of a unitary or anti-unitary operator (Proposition 4.2). So, there is a unitary or anti-unitary operator U such that $L(P) = L_U(P)$ for every projection $P \in \mathcal{F}_s(H)$. This gives the claim (since $\mathcal{F}_s(H)$ is spanned by the set of all finite-rank projections).   □

Theorem 6.7 *Suppose that* $\dim H \geq 3$. *Let L be a linear transformation of* $\mathcal{F}_s(H)$ *such that for a certain natural $k < \dim H$ we have*

$$L(\mathcal{P}_k(H)) \subset \mathcal{P}_k(H)$$

and the restriction of L to $\mathcal{P}_k(H)$ is injective. Then there is a linear or conjugate-linear isometry U such that one of the following possibilities is realized:

- $L = L_U$,
- $\dim H = 2k$ *and* $L = L_k^\perp L_U$.

The proof of Theorem 6.7 is given in Section 6.3.

By Example 6.5, the assumption that the restriction of L to $\mathcal{P}_k(H)$ is injective cannot be omitted.

First of all, we note that the general case can be reduced to the case when $\dim H \geq 2k$. Let L be as in Theorem 6.7. Suppose that $\dim H = n$ is finite and consider the linear transformation $L_k^\perp L L_{n-k}^\perp$ which sends $\mathcal{P}_{n-k}(H)$ to a subset of $\mathcal{P}_{n-k}(H)$. Assume that

$$L_k^\perp L L_{n-k}^\perp = L_U$$

for a certain unitary or anti-unitary operator U. Since $U(X^\perp) = U(X)^\perp$ for every closed subspace $X \subset H$, the transformation L maps the projection on a k-dimensional subspace X to the projection on $U(X)$. Therefore, $L(P) = L_U(P)$ for all $P \in \mathcal{P}_k(H)$ and we have $L = L_U$, since $\mathcal{F}_s(H)$ is spanned by $\mathcal{P}_k(H)$.

Remark 6.8 The above statement was proved by Aniello and Chruściński [1] under the following assumptions:

- $\dim H \neq 2k$,
- L is injective if H is infinite-dimensional,

• the restriction of L to $\mathcal{P}_k(H)$ is a bijective transformation of $\mathcal{P}_k(H)$.

The proof given in [1] is based on the description of bijective transformations of Grassmannians preserving the orthogonality relation in both directions (Theorem 4.10). See [57, 58] for other versions of this result. To prove Theorem 6.7 we will use Theorem 4.26 and some arguments from the proof of Theorem 4.25.

Remark 6.9 Theorem 6.7 has an analogue in linear algebra. If V is a vector space over a field, then the Grassmannian $\mathcal{G}_k(V)$ can be considered as a subset of the projective space associated to the exterior product $\wedge^k V$. There is a description of semilinear transformations of $\wedge^k V$ preserving $\mathcal{G}_k(V)$ [45, Theorem 6.1]. For example, if $\dim V \neq 2k$ and the field is algebraically closed, then every such semilinear transformation is induced by a semilinear automorphism of V (the general case is more complicated). A result of such kind was first proved by Westwick [65].

For two complex Hilbert spaces H and H' we will investigate linear maps of $\mathcal{F}_s(H)$ to $\mathcal{F}_s(H')$ which send $\mathcal{P}_k(H)$ to a subset of $\mathcal{P}_m(H')$. As above, for a linear or conjugate-linear isometry $U : H \to H'$ we denote by L_U the linear injection of $\mathcal{F}_s(H)$ to $\mathcal{F}_s(H')$ induced by U. If $\dim H = \dim H'$ is finite, then U is a unitary or anti-unitary operator and L_U is a linear isomorphism.

Example 6.10 We take two natural numbers $k < \dim H$ and $m < \dim H'$ such that $k \leq m$. Let W be an $(m - k)$-dimensional subspace of H' and let H'' be the orthogonal complement of W. Let also $U : H \to H''$ be a linear or conjugate-linear isometry. For every $A \in \mathcal{F}_s(H)$ we define $L_{U,W}(A)$ as the element of $\mathcal{F}_s(H')$ whose restriction to H'' coincides with $L_U(A)$ and the restriction to W is $k^{-1}\mathrm{tr}(A)\mathrm{Id}_W$, i.e.

$$L_{U,W}(A) = L_U(A)P_{H''} + k^{-1}\mathrm{tr}(A)P_W.$$

Then the map

$$L_{U,W} : \mathcal{F}_s(H) \to \mathcal{F}_s(H')$$

is a linear injection (which coincides with L_U if $W = 0$) and

$$L_{U,W}(\mathcal{P}_k(H)) \subset \mathcal{P}_m(H').$$

It sends the projection on a k-dimensional subspace X to the projection on the m-dimensional subspace $U(X) + W$.

Theorem 6.11 (Pankov [49]) *Let* $\dim H \geq 2k$ *and let* L *be a linear map of* $\mathcal{F}_s(H)$ *to* $\mathcal{F}_s(H')$ *satisfying the following conditions:*

(L1) $L(\mathcal{P}_k(H)) \subset \mathcal{P}_m(H')$ *for some natural k and m,*
(L2) *the restriction of L to $\mathcal{P}_k(H)$ is injective.*

Then $k \le m$. Suppose that H is infinite-dimensional and the following additional condition holds:

(L3) *for any $P, Q \in \mathcal{P}_k(H)$ the dimension of the intersection of the images of $L(P)$ and $L(Q)$ is not less than $m - k$.*

Then $L = L_{U,W}$, where W is an $(m - k)$-dimensional subspace of H' and U is a linear or conjugate-linear isometry of H to the orthogonal complement of W.

The proof of Theorem 6.11 is given in Section 6.3.

It must be pointed out that the condition (L3) holds trivially if $k = m$.

Remark 6.12   Suppose that $\dim H = 2k$ is finite and $k \ge 2$. Let L be a linear map of $\mathcal{F}_s(H)$ to $\mathcal{F}_s(H')$ satisfying (L1) – (L3). In the next section (Remark 6.18), we show that there exist an $(m - k)$-dimensional subspace $W \subset H'$ and a linear or conjugate-linear isometry of H to the orthogonal complement of W such that

$$L = L_{U,W} \quad \text{or} \quad L = L_{U,W} L_k^{\perp}.$$

It is an open problem to obtain an analogue of Theorem 6.11 for the case when $\dim H$ is finite and greater than $2k$. Some aspects of this problem will be described in Remark 6.18.

6.3 Proofs of Theorems 6.7 and 6.11

In this section, we suppose that $\dim H \ge 2k$ and

$$L : \mathcal{F}_s(H) \to \mathcal{F}_s(H')$$

is a linear map satisfying the conditions (L1) and (L2) from Theorem 6.11. Consider the injective map

$$f : \mathcal{G}_k(H) \to \mathcal{G}_m(H')$$

induced by L, i.e. such that

$$L(P_X) = P_{f(X)}$$

for every $X \in \mathcal{G}_k(H)$.

Recall that for any $X, Y \in \mathcal{G}_k(H)$ the set $\mathcal{X}_k(X, Y)$ consists of all $Z \in \mathcal{G}_k(H)$ such that $P_X + P_Y - P_Z$ is a rank-k projection (Section 4.8). The inclusion

$$f(\mathcal{X}_k(X, Y)) \subset \mathcal{X}_m(f(X), f(Y))$$

is obvious.

For any subspaces M and N such that $\dim M \le k \le \dim N$ and $M \subset N$ we denote by $[M, N]_k$ the set formed by all k-dimensional subspaces Z satisfying $M \subset Z \subset N$. By Lemma 4.43, we have

$$\mathcal{X}_k(X, Y) \subset [X \cap Y, X + Y]_k$$

for any $X, Y \in \mathcal{G}_k(H)$.

Lemma 6.13 *The inverse inclusion holds only in the case when X and Y are compatible. In particular, $\mathcal{X}_k(X, Y)$ coincides with $\mathcal{G}_k(X + Y)$ if and only if X, Y are orthogonal.*

Proof It follows from Lemma 4.43 that we can restrict ourselves to the case when $X \cap Y = 0$. Then $\dim(X + Y) = 2k$. If X and Y are orthogonal, then $\mathcal{X}_k(X, Y)$ coincides with $\mathcal{G}_k(X + Y)$ (Example 4.41).

Suppose that $\mathcal{X}_k(X, Y)$ coincides with $\mathcal{G}_k(X + Y)$. Consider a k-dimensional subspace $X' \subset X + Y$ spanned by some eigenvectors of the self-adjoint operator $P_X + P_Y$. Then $P_{X'}$ and $P_X + P_Y$ commute. By our assumption, X' belongs to $\mathcal{X}_k(X, Y)$ and there is a k-dimensional subspace $Y' \subset X + Y$ such that

$$P_X + P_Y = P_{X'} + P_{Y'}. \tag{6.1}$$

Since $P_{X'}$ and $P_X + P_Y$ commute, the latter equality shows that $P_{X'}$ and $P_{Y'}$ commute. Therefore, X' and Y' are compatible. It follows from Lemma 4.42 that $X + Y = X' + Y'$. The dimension of this subspace is $2k$, i.e. X' and Y' are orthogonal. Then (6.1) is a projection, which implies that X and Y are orthogonal (see the proof of Theorem 6.6). $\square$

Now, we modify some arguments exploited to prove Theorem 4.25 (Section 4.8). Suppose that X and Y are orthogonal. Then $\mathcal{G}_k(X + Y) = \mathcal{X}_k(X, Y)$ and

$$f(\mathcal{G}_k(X + Y)) \subset \mathcal{X}_m(f(X), f(Y)) \subset [f(X) \cap f(Y), f(X) + f(Y)]_m. \tag{6.2}$$

Therefore, L transfers $\mathcal{F}_s(X + Y)$ to a subspace of $\mathcal{F}_s(f(X) + f(Y))$ (for every closed subspace $Z \subset H$ the subspace of $\mathcal{F}_s(H)$ formed by all operators whose images are contained in Z can be naturally identified with $\mathcal{F}_s(Z)$). The Grassmannian $\mathcal{G}_k(X + Y)$ is a connected compact topological real manifold of dimension

$$k' = 2k^2.$$

Similarly,

$$[f(X) \cap f(Y), f(X) + f(Y)]_m \tag{6.3}$$

is a connected compact topological real manifold of dimension

$$m' = 2(m - t)^2,$$

where t is the dimension of $f(X) \cap f(Y)$. The vector spaces $\mathcal{F}_s(X + Y)$ and $\mathcal{F}_s(f(X) + f(Y))$ are finite-dimensional and the restriction of L to $\mathcal{F}_s(X + Y)$ is continuous. Since f is induced by L, the restriction of f to $\mathcal{G}_k(X + Y)$ is continuous. By (L2), f is injective.

Lemma 6.14 $m' \geq k'$.

Proof   We choose open subsets U and V in $\mathcal{G}_k(X+Y)$ and (6.3) homeomorphic to $\mathbb{R}^{k'}$ and $\mathbb{R}^{m'}$ (respectively) and such that $f(U) \subset V$. If $k' > m'$, then the restriction of f to U can be considered as a continuous injection

$$f' : \mathbb{R}^{k'} \rightarrow \mathbb{R}^{m'} \subset \mathbb{R}^{k'}$$

whose image is nowhere dense in $\mathbb{R}^{k'}$. On the other hand, the image of every continuous injective transformation of $\mathbb{R}^{k'}$ is an open subset [16, Chapter IV, Proposition 7.4] and we get a contradiction.   □

In particular, Lemma 6.14 shows that $m \geq k$.

From this moment, we assume that L satisfies the condition (L3) from Theorem 6.11, which is equivalent to the fact that

$$\dim(f(X') \cap f(Y')) \geq m - k$$

for any $X', Y' \in \mathcal{G}_k(H)$. Then $m' \leq k'$ and Lemma 6.14 implies that $m' = k'$.

Lemma 6.15   *The following assertions are fulfilled:*

(1) *if $X, Y \in \mathcal{G}_k(H)$ are orthogonal, then $f(X)$ and $f(Y)$ are compatible and*

$$\dim(f(X) \cap f(Y)) = m - k.$$

(2) *f is adjacency preserving in both directions.*

Proof   (1) Suppose that $X, Y \in \mathcal{G}_k(H)$ are orthogonal. The equality $m' = k'$ shows that $f(X) \cap f(Y)$ is $(m - k)$-dimensional. Then $\mathcal{G}_k(X + Y)$ and (6.3) are homeomorphic. Lemma 4.45 guarantees that

$$f(\mathcal{G}_k(X + Y)) = [f(X) \cap f(Y), f(X) + f(Y)]_m$$

and (6.2) shows that

$$\mathcal{X}_m(f(X), f(Y)) = [f(X) \cap f(Y), f(X) + f(Y)]_m.$$

By Lemma 6.13, $f(X)$ and $f(Y)$ are compatible.

(2) For any $X', Y' \in \mathcal{G}_k(H)$ we consider orthogonal $X, Y \in \mathcal{G}_k(H)$ such that

$$X' + Y' \subset X + Y.$$

It was established above that the restriction of f to $\mathcal{G}_k(X + Y)$ is a homeomorphism to (6.3). Therefore, $L(\mathcal{F}_s(X + Y))$ is the subspace of $\mathcal{F}_s(H')$ spanned by all P_Z with Z belonging to (6.3). This subspace can be identified with $\mathcal{F}_s(\tilde{H})$, where $\tilde{H}$ is a complex Hilbert space of dimension $2k$. The restriction of L to $\mathcal{F}_s(X + Y)$ is a linear isomorphism to $\mathcal{F}_s(\tilde{H})$ (as a surjective linear map between vector spaces of the same finite dimension). This means that

$$f(\mathcal{X}_k(X', Y')) = \mathcal{X}_m(f(X'), f(Y')).$$

As in the proof of Theorem 4.25, we establish that f is adjacency preserving in both directions. $\square$

Proof of Theorem 6.7 Suppose that $k = m$ and $H = H'$. Lemma 6.15 states that f is orthogonality preserving and it is also adjacency preserving in both directions (note that the second statement is obvious for $k = 1$).

Let $k = 1$. For every $S \in \mathcal{G}_2(H)$ we take orthogonal $X, Y \in \mathcal{G}_1(S)$. Since $f(X)$ and $f(Y)$ also are orthogonal, we have

$$f(\mathcal{G}_1(S)) = f(\mathcal{X}_1(X, Y)) = \mathcal{X}_1(f(X), f(Y)) = \mathcal{G}_1(f(X) + f(Y)).$$

So, f sends lines to lines. Also, f is injective by (L2) and $f(\mathcal{G}_1(H))$ is not contained in a line. This means that f is induced by a semilinear injective transformation of H. This semilinear transformation is orthogonality preserving (since f is orthogonality preserving), which implies that it is a scalar multiple of a linear or conjugate-linear isometry. Therefore, there is a linear or conjugate-linear isometry U such that $f(X) = U(X)$ for every $X \in \mathcal{G}_1(H)$. Then $L(P) = L_U(P)$ for every $P \in \mathcal{P}_1(H)$ and we obtain that $L = L_U$.

In the case when $k \geq 2$, the statement follows from Theorem 4.26 and Proposition 4.28. $\square$

Now, we suppose that H is infinite-dimensional and prove Theorem 6.11. The case when $k = m$ was considered above and we assume that $m > k$.

Proof of Theorem 6.11 for $k = 1$ If $k = 1$, then $m \geq 2$. The second part of Lemma 6.15 implies that any two distinct elements of $f(\mathcal{G}_1(H))$ are adjacent. Hence $f(\mathcal{G}_1(H))$ is contained in a star or a top of $\mathcal{G}_m(H')$. Consider an infinite set $\mathcal{X}$ formed by mutually orthogonal elements of $\mathcal{G}_1(H)$ (such sets exist, since H is infinite-dimensional). It follows from Lemma 6.15 that $f(\mathcal{X})$ is an infinite set consisting of mutually ortho-adjacent elements of $\mathcal{G}_m(H')$, i.e. $f(\mathcal{X})$ is a compatible subset in a maximal clique of $\Gamma_m(H')$. Since any compatible subset in a top is finite, $f(\mathcal{G}_1(H))$ is contained in a star and there is an $(m-1)$-dimensional subspace W contained in each element of $f(\mathcal{G}_1(H))$.

Let H'' be the orthogonal complement of W. Consider the map

$$g : \mathcal{G}_1(H) \to \mathcal{G}_1(H'')$$

defined as $g(X) = f(X) \cap H''$ for every $X \in \mathcal{G}_1(H)$. This map is orthogonality preserving by Lemma 6.15. For any orthogonal $X, Y \in \mathcal{G}_1(H)$ we have

$$f(\mathcal{G}_1(X + Y)) = [W, f(X) + f(Y)]_m,$$

which implies that g sends lines of Π_H to lines of $\Pi_{H''}$. The map g is injective (since f is injective) and $g(\mathcal{G}_1(H))$ is not contained in a line. Therefore, g is induced by a semilinear injection $S : H \to H''$ and

$$f(X) = S(X) + W$$

for all $X \in \mathcal{G}_1(H)$. Since g is orthogonality preserving, S is a scalar multiple of a linear or conjugate-linear isometry $U : H \to H''$ and we obtain that $L(P) = L_{U,W}(P)$ for all $P \in \mathcal{P}_1(H)$, which gives the claim.   □

Now, let $k \geq 2$. To prove Theorem 6.11 for this case, we modify some arguments used to prove Theorem 4.26.

By the second part of Lemma 6.15, f is adjacency preserving in both directions. We prove the following.

Lemma 6.16 *The map f is ortho-adjacency preserving.*

Proof The proof is similar to the proof of Lemma 4.32. We use the following facts: f is adjacency preserving and for any orthogonal $X, Y \in \mathcal{G}_k(H)$ the subspaces $f(X)$ and $f(Y)$ are compatible and the distance between them in the Grassmann graph $\Gamma_m(H')$ is equal to k.   □

Since f is adjacency preserving in both directions, it transfers every maximal clique of $\Gamma_k(H)$ (a star or a top) to a subset in a maximal clique of $\Gamma_m(H')$ and distinct maximal cliques go to subsets of distinct maximal cliques. By our assumption, H is infinite-dimensional, which guarantees that every star of $\mathcal{G}_k(H)$ contains an infinite subset formed by mutually ortho-adjacent elements. A top of $\mathcal{G}_m(H')$ does not contain such subsets and, by Lemma 6.16, f sends every star $\mathcal{S} \subset \mathcal{G}_k(H)$ to a subset of a star; moreover, $f(\mathcal{S})$ is contained in a unique star of $\mathcal{G}_m(H')$ (since the intersection of two distinct stars contains at most one element). Therefore, f induces an injection

$$f_{k-1} : \mathcal{G}_{k-1}(H) \to \mathcal{G}_{m-1}(H')$$

such that

$$f(S(X)) \subset S(f_{k-1}(X))$$

for every $X \in \mathcal{G}_{k-1}(H)$. Then for every $Y \in \mathcal{G}_k(H)$ we have

$$f_{k-1}(\mathcal{G}_{k-1}(Y)) \subset \mathcal{G}_{m-1}(f(Y)).$$

Since f_{k-1} is injective, the latter inclusion implies that f_{k-1} is adjacency preserving.

Lemma 6.17 *If $X, Y \in \mathcal{G}_{k-1}(H)$ are orthogonal, then $f_{k-1}(X)$ and $f_{k-1}(Y)$ are compatible and*

$$\dim(f_{k-1}(X) \cap f_{k-1}(Y)) = m - k.$$

The map f_{k-1} is ortho-adjacency preserving.

Proof Since f_{k-1} is adjacency preserving, it sends every path of $\Gamma_{k-1}(H)$ to a path of $\Gamma_{m-1}(H')$. Then

$$d(f_{k-1}(X), f_{k-1}(Y)) \le d(X, Y) = k - 1$$

and we have

$$m - 1 - \dim(f_{k-1}(X) \cap f_{k-1}(Y)) = d(f_{k-1}(X), f_{k-1}(Y)) \le k - 1,$$

where d is the distance in Grassmann graphs. Therefore,

$$\dim(f_{k-1}(X) \cap f_{k-1}(Y)) \ge m - k. \tag{6.4}$$

We take orthogonal $X', Y' \in \mathcal{G}_k(H)$ such that $X \subset X'$ and $Y \subset Y'$. Then $f_{k-1}(X)$ and $f_{k-1}(Y)$ are contained in $f(X')$ and $f(Y')$, respectively. Hence

$$f_{k-1}(X) \cap f_{k-1}(Y) \subset f(X') \cap f(Y'),$$

which implies that

$$\dim(f_{k-1}(X) \cap f_{k-1}(Y)) \le \dim(f(X') \cap f(Y')) = m - k.$$

This inequality and (6.4) show that the dimension of $f_{k-1}(X) \cap f_{k-1}(Y)$ is equal to $m - k$ and

$$f_{k-1}(X) \cap f_{k-1}(Y) = f(X') \cap f(Y'). \tag{6.5}$$

The subspaces $f(X')$ and $f(Y')$ are compatible. This implies the existence of orthogonal k-dimensional subspaces $X'' \subset f(X')$ and $Y'' \subset f(Y')$ such that $f(X') \cap f(Y')$ is orthogonal to both X'', Y'' and

$$f(X') = X'' + f(X') \cap f(Y'), \quad f(Y') = Y'' + f(X') \cap f(Y').$$

Since $f_{k-1}(X)$ and $f_{k-1}(Y)$ are hyperplanes in $f(X')$ and $f(Y')$ (respectively),

$$X'' \cap f_{k-1}(X) \quad \text{and} \quad Y'' \cap f_{k-1}(Y)$$

are orthogonal $(k-1)$-dimensional subspaces which are also orthogonal to (6.5). We have

$$f_{k-1}(X) = X'' \cap f_{k-1}(X) + f_{k-1}(X) \cap f_{k-1}(Y),$$

$$f_{k-1}(Y) = Y'' \cap f_{k-1}(Y) + f_{k-1}(X) \cap f_{k-1}(Y).$$

So, $f_{k-1}(X)$ and $f_{k-1}(Y)$ are compatible.

As in the proof of Lemma 6.16, we modify the arguments from the proof of Lemma 4.32 and establish that f_{k-1} is ortho-adjacency preserving.   □

Suppose that $k \geq 3$. Using the fact that f_{k-1} is adjacency and ortho-adjacency preserving, we show that for every star $\mathcal{S} \subset \mathcal{G}_{k-1}(H)$ there is a unique star of $\mathcal{G}_{m-1}(H')$ containing $f_{k-1}(\mathcal{S})$. Therefore, f_{k-1} induces a map

$$f_{k-2} : \mathcal{G}_{k-2}(H) \to \mathcal{G}_{m-2}(H')$$

which transfers tops to subsets of tops. We cannot state that f_{k-2} is injective (since f_{k-1} is not necessarily adjacency preserving in both directions). For any adjacent $X, Y \in \mathcal{G}_{k-2}(H)$ the images $f_{k-2}(X)$ and $f_{k-2}(Y)$ are adjacent or coincident. This means that f_{k-2} sends any path of $\Gamma_{k-2}(H)$ to a path of $\Gamma_{m-2}(H')$ (possibly of shorter length). We repeat the above arguments and obtain the direct analogue of Lemma 6.17 for f_{k-2}.

Proof of Theorem 6.11 for $k \geq 2$ Recursively, we construct a sequence of maps

$$f_i : \mathcal{G}_i(H) \to \mathcal{G}_{m-k+i}(H')$$

such that $f_k = f$ and

$$f_i(\mathcal{S}(X)) \subset \mathcal{S}(f_{i-1}(X))$$

for every $X \in \mathcal{G}_{i-1}(H)$ and $i \geq 2$. Then for every $Y \in \mathcal{G}_{i+1}(H)$ and $i \leq k-1$ we have

$$f_i(\mathcal{G}_i(Y)) \subset \mathcal{G}_{m-k+i}(f_{i+1}(Y)).$$

We will exploit the following properties of f_1:

(A) f_1 transfers lines to subsets in tops of $\mathcal{G}_{m-k+1}(H')$ (f_1 is induced by f_2),
(B) f_1 sends orthogonal elements to ortho-adjacent elements.

By (A), for any distinct $X, Y \in \mathcal{G}_1(H)$ the images $f_1(X), f_1(Y)$ are adjacent or coincident. This means that $f_1(\mathcal{G}_1(H))$ is contained in a maximal clique of $\Gamma_{m-k+1}(H')$. Using (B), we show that this maximal clique is a star (see the proof of Theorem 6.11 for $k = 1$). So, there is an $(m - k)$-dimensional subspace W contained in every element of $f_1(\mathcal{G}_1(H))$.

The inclusion

$$f_1(\mathcal{G}_1(X)) \subset \mathcal{G}_{m-k+1}(f(X)) \text{ for all } X \in \mathcal{G}_k(H) \qquad (6.6)$$

follows easily from the fact that f_i is induced by f_{i+1} for all $i \leq k - 1$. Next, we claim that the map f_1 is injective (the proof is similar to the proof of Lemma 4.36).

As in the proof of Theorem 6.11 for $k = 1$, we denote by H'' the orthogonal complement of W and consider the map $g : \mathcal{G}_1(H) \to \mathcal{G}_1(H'')$ defined as

$$g(X) = f_1(X) \cap H''$$

for every $X \in \mathcal{G}_1(H)$. By (B), this map is orthogonality preserving; in particular, the image $g(\mathcal{G}_1(H))$ is not contained in a line of Π''_H. The map g is injective (since f_1 is injective) and it sends lines to subsets of lines by (A). Therefore, there is a linear or conjugate-linear isometry $U : H \to H''$ such that

$$f_1(X) = U(X) + W$$

for all $X \in \mathcal{G}_1(H)$. Using (6.6), we show that

$$f(X) = U(X) + W$$

for each $X \in \mathcal{G}_k(H)$. Then $L(P) = L_{U,W}(P)$ for all $P \in \mathcal{P}_k(H)$. □

Remark 6.18 Consider the case when H is of an arbitrary (not necessarily infinite) dimension not less than three and $\dim H \geq 2k$. As above, we assume that f is the injection of $\mathcal{G}_k(H)$ to $\mathcal{G}_m(H')$ induced by the linear map L. Then f is adjacency preserving in both directions (Lemma 6.15) and it also preserves the ortho-adjacency relation (the assumption that H is infinite-dimensional is not exploited to prove Lemma 6.16). The first property guarantees that one of the following possibilities is realized:

(S) f sends all stars to subsets of stars,

(T) all stars go to subsets of tops

(see the proof of Proposition 4.28). Using the second property, we establish that (S) holds for the infinite-dimensional case. Now, we suppose that $\dim H = n$ is finite. The arguments from the proof of Theorem 6.11 work only for the case (S). The cardinalities of maximal compatible subsets in stars of $\mathcal{G}_k(H)$ and tops of $\mathcal{G}_m(H')$ are $n - k + 1$ and $m + 1$, respectively. Therefore, if $m \geq n - k$, then we cannot assert that the possibility (T) is not realized. If $n = 2k$, then we take any orthogonal $X, Y \in \mathcal{G}_k(H)$ and establish that f is a homeomorphism of $\mathcal{G}_k(H) = \mathcal{X}_k(X, Y)$ to

$$\mathcal{X}_m(f(X), f(Y)) = [W, N]_m,$$

where W and N are $(m - k)$-dimensional and $(m + k)$-dimensional subspaces of H'. In this case, we reduce L to a linear map of $\mathcal{F}_s(H)$ to $\mathcal{F}_s(\tilde{H})$ sending $\mathcal{P}_k(H)$ to $\mathcal{P}_k(\tilde{H})$, where $\tilde{H}$ is $2k$-dimensional; next, we apply Theorem 6.7. If $n > 2k$, then for every $2k$-dimensional subspace $X \subset H$ there is an $(m - k)$-dimensional subspace $W_X \subset H'$ and an $(m + k)$-dimensional subspace $N_X \subset H'$ such that

$$f(\mathcal{G}_k(X)) = [W_X, N_X]_m.$$

Is it true that all W_X are coincident? Equivalently, is there an $(m-k)$-dimensional subspace $W \subset H'$ contained in all elements of the image of f? The existence of such W can be established only in the case (S).

References

[1] P. Aniello, D. Chruściński, *Symmetry witnesses*, J. Phys. A, Math. Theor. 50(2017), No. 28, 16 pp.

[2] B.H. Arnold, *Rings of operators on vector spaces*, Ann. of Math. 45(1944), 24–49.

[3] R. Baer, *Linear Algebra and Projective Geometry*, Academic Press, 1952.

[4] J. Barvinek, J. Hamhalter, *Linear algebraic proof of Wigner theorem and its consequences*, Math. Slovaca 67(2017), 371–386.

[5] R. Bhatia, *Matrix Analysis*, Springer, 1997.

[6] G. Birkhoff, J. von Neumann, *The logic of quantum mechanics*, Ann. of Math. 36(1937), 823–843.

[7] A. Blunck, H. Havlicek, *The connected components of the projective line over a ring*, Adv. Geom. 1(2001), 107–117.

[8] A. Blunck, H. Havlicek, *On bijections that preserve complementarity of subspaces*, Discrete Math. 301(2005), 46–56.

[9] F. Botelho, J. Jamison, L. Molnár, *Surjective isometries on Grassmann spaces*, J. Funct. Anal. 265(2013), 2226–2238.

[10] A. Böttcher, I.M. Spitkovsky, *A gentle guide to the basics of two projections theory*, Linear Algebra Appl. 432(2010), 1412–1459.

[11] G. Cassinelli, E. De Vito, P.J. Lahti, A. Levrero, *The Theory of Symmetry Actions in Quantum Mechanics with Applications to the Galilei Group*, Springer, 2004.

[12] G. Chevalier, *Wigner's theorem and its generalizations*, Handbook of Quantum Logic and Quantum Structures, pp. 429–475, Elsevier, 2007.

[13] W.L. Chow, *On the geometry of algebraic homogeneous spaces*, Ann. of Math. 50(1949), 32–67.

[14] D.W. Cohen, *An Introduction to Hilbert Space and Quantum Logic*, Problem Books in Mathematics, Springer, 1989.

[15] J. Dieudonné, *La Géométrie des Groupes Classiques*, Springer, 1971.

[16] A. Dold, *Lectures on Algebraic Topology*, Classics in Mathematics, 2nd ed., Springer, 1980.

[17] F.R. Drake, *Set Theory: An Introduction to Large Cardinals*, Studies in Logic and the Foundations of Mathematics 76, Elsevier, 1974.

[18] J. Duncan, P.J. Taylor, *Norm inequalities for C^*-algebras*, Proc. Roy. Soc. Edinburgh 75(1976), 119–129.

[19] A. Dvurečenskij, *Gleason's Theorem and Its Applications*. Kluwer Academic, 1993.

[20] M. Eidelheit, *On isomorphisms of rings of linear operators*, Studia Math. 9 (1940), 97–105.

[21] C.A. Faure, A. Frölicher, *Morphisms of projective geometries and semilinear maps*, Geom. Dedicata 53(1994), 237–262.

[22] P.A. Fillmore, W.E. Longstaff, *On isomorphisms of lattices of closed subspaces*, Canad. J. Math. 36(1984), 820–829.

[23] G.P. Gehér, *An elementary proof for the non-bijective version of Wigner's theorem*, Phys. Lett. A 378(2014), 2054–2057.

[24] G.P. Gehér, P. Šemrl, *Isometries of Grassmann spaces*, J. Funct. Anal. 270(2016), 1585–1601.

[25] G.P. Gehér, *Wigner's theorem on Grassmann spaces*, J. Funct. Anal. 273(2017), 2994–3001.

[26] A.M. Gleason, *Measures on the closed subspaces of a Hilbert space*, Indiana Univ. Math. J. 6(1957), 885–893.

[27] M. Györy, *Transformations on the set of all n-dimensional subspaces of a Hilbert space preserving orthogonality*, Publ. Math. Debrecen 65(2004), 233–242.

[28] H. Havlicek, *A generalization of Brauner's theorem on linear mappings*, Mitt. Math. Sem. Univ. Giessen 215(1994), 27–41.

[29] C. Jordan, *Essai sur la géométrie á n dimensions*, Bull. Soc. Math. France 3(1875), 103–174.

[30] R.V. Kadison, *Transformations of states in operator theory and dynamics*, Topology 3(1965), 177–198.

[31] S. Kakutani, G.W. Mackey, *Ring and lattice characterizations of complex Hilbert space*, Bull. Amer. Math. Soc. 52(1946), 727–733.

[32] T. Kato, *Perturbation Theory for Linear Operators*, Classics in Mathematics, 2nd ed., Springer, 1995.

[33] A. V. Knyazev, A. Jujunashvili, M. Argentati, *Angles between infinite dimensional subspaces with applications to the Rayleigh–Ritz and alternating projectors methods*, J. Funct. Anal. 259(2010), 1323–1345.

[34] S. Lang, *Algebra*, Graduate Texts in Mathematics 211, Springer, 2002.

[35] G.W. Mackey, *Isomorphisms of normed linear spaces*, Ann. of Math. 43(1942), 244–260.

[36] L. Molnár, *Transformations on the set of all n-dimensional subspaces of a Hilbert space preserving principal angles*, Comm. Math. Phys. 217(2001), 409–421.

[37] L. Molnár, *Selected Preserver Problems on Algebraic Structures of Linear Operators and on Function Spaces*, Lecture Notes in Mathematics 1895, Springer, 2007.

[38] L. Molnár, *Maps on the n-dimensional subspaces of a Hilbert space preserving principal angles*, Proc. Amer. Math. Soc. 136(2008), 3205–3209.

[39] L. Molnár, W. Timmermann, *A metric on the space of projections admitting nice isometries*, Studia Math. 191(2009), 271–281.

[40] L. Molnár, P. Šemrl, *Transformations of the unitary group of a Hilbert space*, J. Math. Anal. Appl. 388(2012), 1205–1217.

[41] P.G. Ovchinikov, *Automorphisms of the poset of skew projections*, J. Funct. Anal. 115(1993), 184–189.

[42] Pankov M., *Order preserving transformations of the Hilbert Grassmannian*, Arch. Math. (Basel) 89(2007), 81–86.

[43] Pankov M., *Order preserving transformations of the Hilbert Grassmannian (note on the complex case)*, Arch. Math. (Basel) 90(2008), 528–529.

[44] Pankov M., *Grassmannians of Classical Buildings*, Algebra and Discrete Mathematics 2, World Scientific, 2010.

[45] Pankov M., *Geometry of Semilinear Embeddings: Relations to Graphs and Codes*, World Scientific, 2015.

[46] Pankov M., *Orthogonal apartments in Hilbert Grassmannians*, Linear Algebra Appl. 506(2016), 168–182.

[47] M. Pankov, *Apartments preserving transformations of Grassmannians of infinite-dimensional vector spaces*, Linear Algebra Appl. 531(2017), 498–509.

[48] M. Pankov, *Geometric version of Wigner's theorem for Hilbert Grassmannians*, J. Math. Anal. Appl. 459(2018), 135–144.

[49] M. Pankov, *Wigner's type theorem in terms of linear operators which send projections of a fixed rank to projections of other fixed rank*, J. Math. Anal. Appl. 474(2019), 1238–1249 .

[50] K.M. Parthasarathy, *Mathematical Foundation of Quantum Mechanics*, Texts and Readings in Mathematics 35, Hindustan Book Agency, 2005.

[51] L. Plevnik, *Maps on essentially infinite idempotent operators*, J. Math. Anal. Appl. 387(2012), 24–32.

[52] L. Plevnik, *Top stars and isomorphisms of Grassmann graphs*, Beiträge Algebra Geom. 56(2015), 703–728.

[53] L. Qiu, Y. Zhang, C.K. Li, *Unitarily invariant metrics on the Grassmann space.* SIAM J. Matrix Anal. Appl. 27(2005), 507–531.

[54] C.E. Rickart, *Isomorphic groups of linear transformations*, Amer. J. Math., 72(1950), 451–464.

[55] W. Rudin, *Functional Analysis*, 2nd ed., McGraw-Hill, 1991.

[56] E. Santos, *The Bell inequalities as tests of quantum logics*, Phys. Lett. A 115(1986), 363–365.

[57] G. Sarbicki, D. Chruściński, M. Mozrzymas, *Generalising Wigner's theorem*, J. Phys. A, Math. Theor. 49(2016), No. 30, 7 pp.

[58] E. Størmer, *Positive maps which map the set of rank k projections onto itself*, Positivity 21(2017), 509–511.

[59] P. Šemrl, *Orthogonality preserving transformations on the set of n-dimensional subspaces of a Hilbert space*, Illinois J. Math. 48(2004), 567–573.

[60] S. Sternberg, *Lectures on Differential Geometry*, Chelsea, 1963.

[61] J. Tits, *Buildings of Spherical Type and Finite BN-Pairs*, Lecture Notes in Mathematics 386, Springer, 1974.

[62] U. Uhlhorn, *Representation of symmetry transformations in quantum mechanics*, Ark. Fys. 23(1963), 307–340.

[63] V.S. Varadarajan, *Geometry of Quantum Theory*, 2nd ed., Springer, 2000.

[64] I. Vidav, *The norm of the sum of two projections*, Proc. Amer. Math. Soc. 65(1977), 297–298.

[65] R. Westwick, *Linear transformations of Grassmann spaces*, Pacific J. Math. 14(1964), 1123–1127.

[66] R. Westwick, *On adjacency preserving maps*, Canad. Math. Bull. 17(1974), 403–405.

[67] E.P. Wigner, *Gruppentheorie und ihre Anwendung auf die Quantenmechanik der Atomspektrum*, Fredrik Vieweg und Sohn, 1931 (English translation *Group Theory and its Applications to the Quantum Mechanics of Atomic Spectra*, Academic Press, 1959).

Index